NMR: the toolkit

P. J. Hore

Reader in Physical Chemistry, Physical and Theoretical Chemistry Laboratory and Corpus Christi College, Oxford University

J. A. Jones

Royal Society University Research Fellow, Clarendon Laboratory and Corpus Christi College, Oxford University

S. Wimperis

Reader in Magnetic Resonance, School of Chemistry, Exeter University

Series sponsor: AstraZeneca

AstraZeneca is one of the world's leading pharmaceutical companies with a strong research base. Its skill and innovative ideas in organic chemistry and bioscience create products designed to fight disease in seven key therapeutic areas: cancer, cardiovascular, central nervous system, gastrointestinal, infection, pain control, and respiratory.

AstraZeneca was formed through the merger of Astra AB of Sweden and Zeneca Group PLC of the UK. The company is headquartered in the UK with over 50,000 employees worldwide. R&D centres of excellence are in Sweden, the UK, and USA with R&D headquarters in Södertälje, Sweden.

AstraZeneca is committed to the support of education in chemistry and chemical engineering.

OXFORD
UNIVERSITY PRESS

OXFORD
UNIVERSITY PRESS

Great Clarendon Street, Oxford OX2 6DP
Oxford University Press is a department of the University of Oxford.
It furthers the University's objective of excellence in research, scholarship,
and education by publishing worldwide in

Oxford New York
Athens Auckland Bangkok Bogotá Buenos Aires Calcutta
Cape Town Chennai Dar es Salaam Delhi Florence Hong Kong Istanbul
Karachi Kuala Lumpur Madrid Melbourne Mexico City Mumbai
Nairobi Paris São Paolo Singapore Taipei Tokyo Toronto Warsaw

with associated companies in Berlin Ibadan

Oxford is a registered trade mark of Oxford University Press
in the UK and in certain other countries

Published in the United States
by Oxford University Press Inc., New York

First published 2000

A catalogue record for this book is available from the British Library

Library of Congress Cataloging in Publication Data
(Data applied for)

ISBN 0 19 8504152

Typeset by the authors
Printed in Great Britain
on acid-free paper by Bath Press Ltd, Bath, Avon

Series Editor's Foreword

Oxford Chemistry Primers are designed to provide clear and concise introductions to a wide range of topics that may be encountered by chemistry students as they progress from the freshman stage through to graduation. The Physical Chemistry series contains books easily recognised as relating to established fundamental core material that all chemists will need to know, as well as books reflecting new directions and research trends in the subject, thereby anticipating (and perhaps encouraging) the evolution of modern undergraduate courses.

In this Physical Chemistry Primer Peter Hore, Jonathan Jones and Stephen Wimperis provide a beautifully written and elegantly presented account of how modern *Nuclear Magnetic Resonance* experiments work. The book develops the internationally acclaimed and best-selling introductory primer on the subject written by Peter Hore and explains in accessible terms the conceptual and theoretical tools which are vital for all chemists wishing to have an appreciation of the full power of this most versatile and informative branch of spectroscopy.

This Primer will be of interest to all students of chemistry (and their mentors).

Richard G. Compton
Physical and Theoretical Chemistry Laboratory,
University of Oxford

Preface

The words 'bewildering' and 'baffling' are frequently found in reviews of books on nuclear magnetic resonance (NMR). They usually refer not to the texts themselves but to the vast, perplexing and ever-expanding array of techniques, applications, pulse sequences, jargon and, of course, acronyms that characterize modern NMR spectroscopy. To understand and use NMR these days, it is no longer sufficient to be adept at interpreting chemical shifts, spin–spin couplings and multiplet patterns. One must be comfortable in two, three or four frequency dimensions, relaxed about experiments on heteronuclear spin systems, excited by multiple-quantum coherence and in tune with sequences of radiofrequency pulses that resemble the sheet music of a Beethoven sonata. This book attempts to explain how some of these experiments work.

Oxford and Exeter
March 2000

P. J. H.
J. A. J.
S. W.

Contents

Part B

Preamble

Not another NMR book? Well, yes and no. There are many excellent NMR texts on the market written for everyone from the neophyte to the connoisseur; we hope this one will be a bit different. It is intended as a short, approachable description of how modern NMR experiments work, aimed principally at those who use, or might use, an NMR spectrometer and are curious about why the spectra look the way they do. We say little about how to perform the experiments, nor do we discuss applications, both of which are well documented elsewhere. What we hope to do is to provide, in an accessible and relatively informal way, the conceptual and theoretical tools needed to understand the inner workings of some of the more important multi-pulse, multi-nuclear, multi-dimensional techniques that chemists and biochemists use to probe the structures and dynamics of molecules in liquids. There is no attempt at a comprehensive coverage.

In a sense this is two books in one, going over similar ground in different ways. In principle, one could read either independently of the other, although it would probably be wiser to begin with Chapter 1. Part A (Chapters 1–6) starts with the *vector model*, a pictorial description of simple NMR experiments, and proceeds to the more powerful *product operator formalism* with which one can appreciate the mechanics of many complex pulse sequences and predict the kinds of spectra they produce. After discussing some quite sophisticated experiments towards the end of Part A, we go back to basics in Part B (Chapters 7–10) and show how straightforward *quantum mechanics* can be used to understand NMR at a more fundamental level. Among other things, Part B attempts to show what product operators really are and to provide justifications for some of the ideas and results the reader was asked to take on trust in Part A. It will also show how one can handle experiments that are beyond the scope of product operators.

We assume the reader is broadly familiar with the fundamental interactions that control the appearance of simple liquid-state NMR spectra, namely chemical shifts and spin–spin couplings. Chapters 1–3 of P. J. Hore's Oxford Chemistry Primer on NMR may provide a useful background; a quick glance through the parts of Chapter 6 that introduce the vector model and two-dimensional NMR might also be helpful.

We refer to this book as OCP 32.

Although the treatment is of necessity mathematical, we have tried to keep things as simple as is consistent with a reasonable level of accuracy. The margins carry brief reminders of bits and pieces of algebra, which the more mathematically sophisticated will probably wish to ignore. There are also a few appendices containing material that would disrupt the flow of the text; these can usually be omitted on a first reading or ignored altogether.

1 The vector model

1.1 Introduction

An introductory account of the vector model can be found in Section 6.2 of OCP 32.

The so-called *vector model* of NMR spectroscopy is an essential weapon in the armoury of every practising NMR spectroscopist because it provides the sort of simple, intuitive, non-mathematical picture that every human brain eventually requires. Although it is an excellent way of understanding NMR experiments on *isolated* spin-1/2 nuclei, the vector model has only limited applicability for interacting or 'coupled' spins, and cannot be used at all to appreciate the inner workings of many important NMR experiments. Nevertheless, it is fundamental to NMR and forms the basis for a much more versatile and powerful formalism, the *product operator* description, which will be introduced in Chapters 3–5. We assume the reader has some acquaintance with the vector model, which we now review.

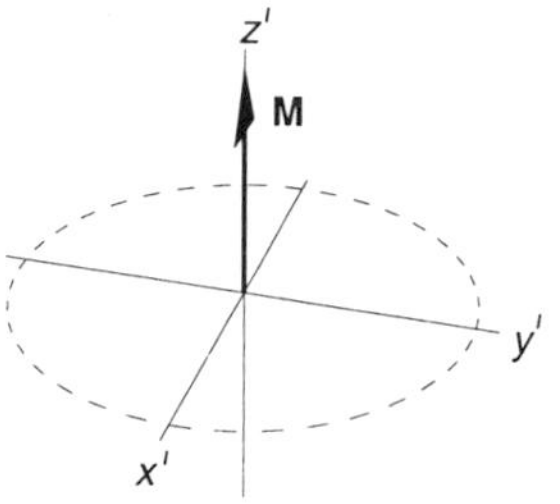

Fig. 1.1 The bulk magnetization **M** in the laboratory frame at thermal equilibrium.

1.2 Bulk magnetization

Elementary quantum mechanics reveals that a nucleus with spin quantum number I in a magnetic field has $2I + 1$ non-degenerate energy levels. For example, a ^1H nucleus (or proton) has spin $I = 1/2$ and is found to occupy one of two distinct energy levels, often labelled α and β. The classical (but incorrect) picture of this situation is that the nuclear magnetic moment precesses in the magnetic field $\mathbf{B_0}$ and has its precession axis aligned either parallel (low energy) or antiparallel (high energy) to $\mathbf{B_0}$. In a macroscopic sample at thermal equilibrium there will be a very slight excess of spins in the lower energy level which leads to a net or *bulk magnetization* **M** of the sample; **M** is stationary and aligned parallel to $\mathbf{B_0}$ (Fig. 1.1). The direction of the magnetic field $\mathbf{B_0}$ defines the z' axis of an x', y', z' coordinate system, which is known as the *laboratory frame*; the x' and y' axes are (arbitrarily) fixed in space. (The reason for putting 'primes' on x', y' and z' will soon become clear.)

We use **bold face** here to denote vector quantities, which have both an amplitude and direction. Often the same symbol in italic type is used for the length of the vector. Thus, B_0 denotes the strength of the magnetic field $\mathbf{B_0}$.

1.3 The rotating frame

Ignoring chemical shifts and J couplings (also known as scalar or spin–spin couplings), the Larmor frequency ω_0 equals $-\gamma B_0$.

A *pulse* is a linearly oscillating magnetic field applied along (say) the x' axis of the laboratory frame. The frequency of its oscillation ω_{rf} (the *transmitter frequency*) is very close to the *Larmor*, or resonance, frequency ω_0 of the spins (typically several hundred megahertz); hence it is usually called the *radiofrequency field*. The effect of a pulse is to tilt the magnetization vector **M** away from the z' axis. Once this motion starts, however, **M** will also immediately begin to precess about the z' axis (i.e. $\mathbf{B_0}$) at the Larmor frequency. These superimposed motions are difficult to visualize because of their rapid and complicated time dependence. The problem can be simplified by thinking of the linearly oscillating radiofrequency field as the sum of two counter-rotating fields with angular frequencies $+\omega_{rf}$ and $-\omega_{rf}$. Only the component that rotates in the same sense as the Larmor precession ($+\omega_{rf}$) is retained; the other ($-\omega_{rf}$) is hundreds of MHz off-resonance and has little effect on the spins. If the NMR experiment is now viewed in a *rotating frame*, which rotates about the z' axis with angular frequency $+\omega_{rf}$, the rotating radiofrequency field appears to be static. Thus in the rotating frame one can view a pulse simply as the temporary application of a static magnetic field $\mathbf{B}_1$, orthogonal to $\mathbf{B}_0$. The other consequence of moving into the rotating frame is that the static field $\mathbf{B}_0$ is replaced by an *effective*, or *residual*, field $\Delta\mathbf{B}_0$ along the z' axis (see Section 1.5). In many situations, the radiofrequency field $\mathbf{B}_1$, when present, is strong enough that $\Delta\mathbf{B}_0$ can be ignored. The axes of the rotating frame are simply labelled x, y, and z with the last corresponding to the z' axis of the laboratory frame.

1.4 Pulses

In the rotating frame, the effect of a radiofrequency pulse is simple: once the pulse is turned on, the bulk magnetization 'sees' an apparently static magnetic field $\mathbf{B}_1$ and precesses (or *nutates*) about it until it is turned off (Fig. 1.2).

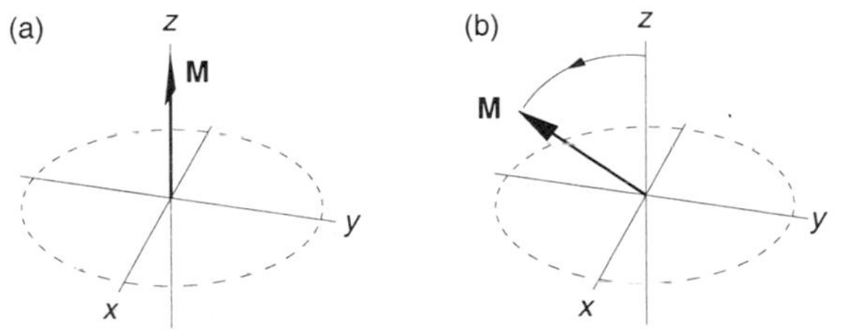

Fig. 1.2 The effect of a radiofrequency pulse. (a) Before the pulse, **M** is aligned along the z axis of the rotating frame. (b) During the pulse, **M** precesses about the $\mathbf{B}_1$ field along the x axis.

Note that, contrary to what is found in many elementary texts (including OCP 32), the axes and the sense of precession have been defined such that a pulse about the $+x$ axis takes **M** initially towards the $-y$ axis. The rate or angular frequency of precession (the *nutation frequency*) is given by $\omega_1 = -\gamma B_1$, where γ is the gyromagnetic ratio of the nuclei being pulsed. The angle β through which **M** nutates is given by $\beta = \omega_1 t_p$, where t_p is the *pulse length*. This time can be chosen to make the *flip angle* β equal to $\pi/2$ (a 90° pulse) if maximum excitation of transverse magnetization is required, or π (a 180° pulse) to invert the equilibrium magnetization (Fig. 1.3).

It does not really matter which sign convention is adopted, provided that everything is done consistently. The different conventions account for the minor discrepancies between equations and figures here and in Section 6.2 of OCP 32.

Most modern spectrometers can apply pulses about any axis in the xy plane of the rotating frame. In many experiments, however, only pulses about the x, y, $-x$ and $-y$ axes are used. This shifting of the direction of the pulse axis in the rotating frame is achieved by altering the *phase* of the radiofrequency field in the laboratory frame and not, of course, by any physical movement of the radiofrequency coil. It should also be remembered that phase is not an absolute term: when we talk of the phase of a pulse we actually mean the phase of the radiofrequency field relative to the phase of the reference frequency of the detector (see Chapter 2).

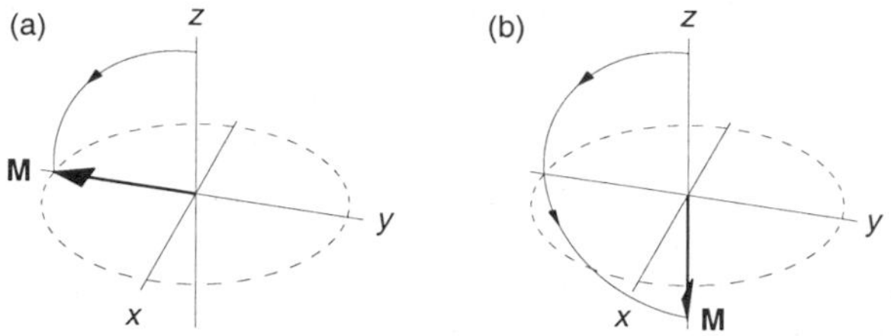

Fig. 1.3 The bulk magnetization **M**, (a) after a 90° pulse and (b) after a 180° pulse, both about the x axis in the rotating frame.

1.5 Free precession

If the NMR spectrum consists of a single peak, for example the ^1H spectrum of water, then the transmitter frequency can be set precisely equal to the Larmor frequency, $\omega_{rf} = \omega_0$. A $90°_x$ pulse rotates the bulk magnetization **M** to the $-y$ axis, but it does *not* then start to precess about the z axis. This is because the transformation from the laboratory frame to the rotating frame has, in this case, completely removed the static magnetic field $\mathbf{B_0}$. The residual field ΔB_0 along the z axis in the rotating frame is $\Delta B_0 = B_0 + \omega_{rf}/\gamma = -(\omega_0 - \omega_{rf})/\gamma$ which vanishes if $\omega_{rf} = \omega_0$, i.e. if the pulse is 'on resonance'.

If the *resonance offset* $\Omega = \omega_0 - \omega_{rf}$ is non-zero then there will be a residual field $\Delta B_0 = -\Omega/\gamma$ in the rotating frame and **M** will start to precess about it as soon as the pulse is switched off. The sense and rate of precession are governed by the sign and magnitude of Ω (Fig. 1.4). The angle through which **M** precesses in the rotating frame in a time t is Ωt radians.

The origin of the residual field in the rotating frame is covered in Appendix 8.4.

ω_0, and therefore Ω, depend on the chemical environment of the nucleus via its chemical shift and J couplings.

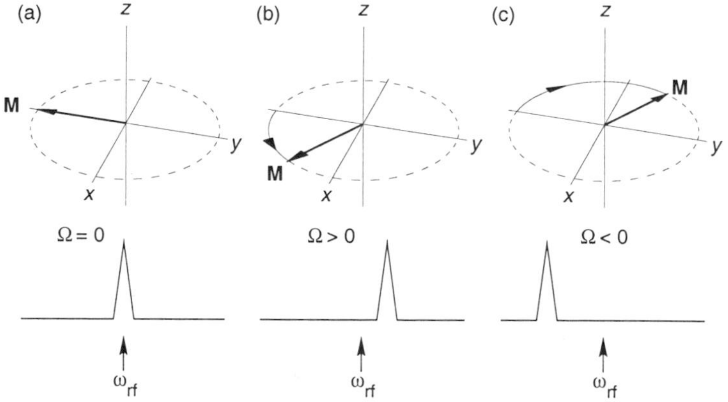

Fig. 1.4 The sense and rate of precession of **M** in the rotating frame depend on the magnitude and sign of the resonance offset Ω: (a) $\Omega = 0$ (on resonance); (b) $\Omega > 0$; (c) $\Omega < 0$. Schematic NMR spectra are shown in the lower part of the figure.

1.6 T_1 and T_2 relaxation

The magnetization **M** produced, for example, by a 90° pulse does not precess in the *xy* plane indefinitely. The populations of the energy levels, which are *equal* immediately after a 90° pulse, typically return to thermal equilibrium—the Boltzmann distribution—in a few seconds (although in some samples it can take microseconds, milliseconds, minutes or even hours). This process is known as *spin–lattice* or *longitudinal relaxation*. Very often it is approximately exponential and can be characterized by a single time constant, T_1. Longitudinal relaxation is usually measured with the *inversion recovery* method: a 180° pulse aligns **M** along the $-z$ axis and then, after a variable time, t, a 90° pulse monitors the amount of magnetization present. By repeating the experiment for a range of values of t, the T_1 time constant can be determined.

Chapter 5 of OCP 32 discusses spin relaxation in more detail.

Section 6.3 of OCP 32 gives a more complete description of the inversion recovery experiment.

Magnetization in the *xy* plane (transverse magnetization) created by a pulse also decays with time, but in this case back to its equilibrium value of zero. This process, called *spin–spin* or *transverse relaxation*, is often described as a loss of *phase coherence* between the individual spins. Usually it is approximately exponential and can be characterized by a single time constant, T_2. In many experiments, the pulses are so short that one can ignore relaxation while the radiofrequency field is present.

1.7 Spin echoes

The vector model is ideal for explaining the classic NMR phenomenon of the *spin echo*. Imagine an NMR spectrum consisting of a single peak, for example the ^{1}H spectrum of water. A $90°_x$ pulse aligns **M** along the $-y$ axis (Fig. 1.5). The magnetization then precesses freely for a time t before a $180°_y$ pulse flips **M** to its mirror image position on the other side of the y axis. Free precession for an identical period t then causes **M** once again to be aligned along the $-y$ axis, irrespective of the resonance offset Ω. The 180° pulse is said to *refocus* the chemical shift or resonance offset of the water, and is often referred to as a *refocusing* pulse. If there are many resonances in the spectrum, all their chemical shifts will be refocused along the same axis by the spin echo.

See also Section 6.3 of OCP 32.

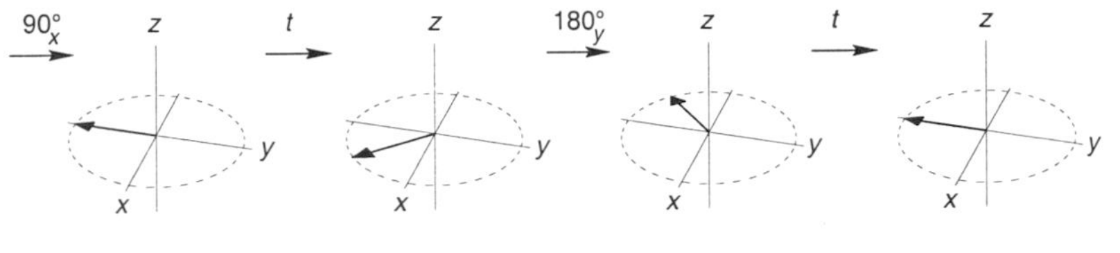

Fig. 1.5 A spin echo for an isolated nucleus (no *J* coupling). The chemical shift/resonance offset is refocused after the second period of free precession.

But what happens when the NMR sample contains a pair of interacting (*J* coupled) spins? The answer depends on whether the two nuclei are of the same species (*homonuclear*, e.g. two protons) or not (*heteronuclear*, e.g. ^{1}H and ^{13}C).

The spin–spin coupling constant J is always quoted in Hertz rather than radians per second, hence the 2π.

Let us start with the easier heteronuclear case and focus on what happens to the magnetization of a ^{1}H nucleus coupled to a ^{13}C. The normal ^{1}H NMR spectrum is a doublet, centred at the ^{1}H resonance offset Ω with splitting $2\pi J$. The two components of the doublet correspond to the α and β spin orientations of the ^{13}C coupling partner. There are therefore two ^{1}H magnetization vectors, with precession frequencies $\Omega \pm \pi J$. A spin echo experiment, with radiofrequency pulses at the ^{1}H NMR frequency, leads to complete refocusing of the heteronuclear J coupling, as well as the offset as shown in Fig. 1.6.

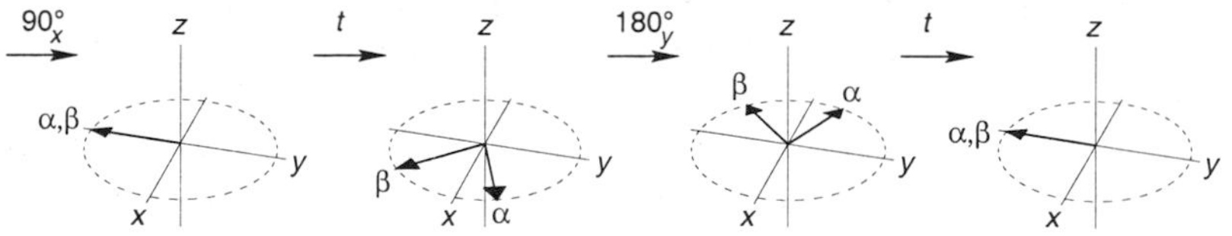

Fig. 1.6 A spin echo for a nucleus with *heteronuclear* J coupling to an $I = 1/2$ partner. The two magnetization vectors represent the components of the NMR doublet; α and β denote the spin orientations of the coupling partner. Both the chemical shift and the J coupling are refocused at time $2t$.

In contrast, homonuclear J couplings are *not* refocused because the α and β spin states of the coupling partner are exchanged by the 180° pulse at the same time as the two magnetization vectors are flipped across the xy plane (Fig. 1.7). The angle between the two vectors continues to grow, and at the end of the $2t$ period is $(2\pi J) \times (2t) = 4\pi Jt$. The echo amplitude is said to be *modulated* by the J coupling: the two vectors are returned to the $-y$ axis if $t = 1/J$.

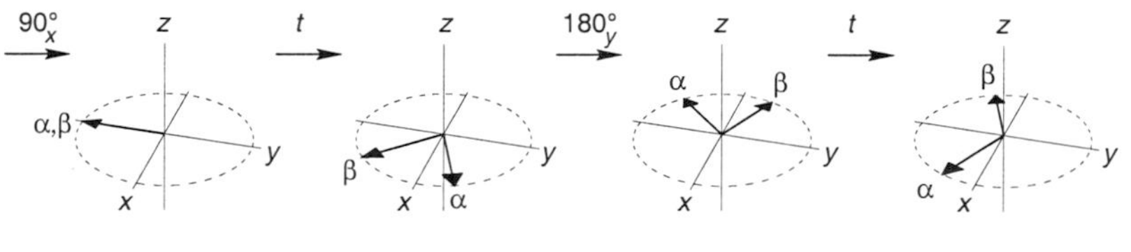

Fig. 1.7 A spin echo for a nucleus with homonuclear J coupling to an $I = 1/2$ partner. Now the J coupling is not refocused because the 180° pulse also inverts the partner spin and interchanges the positions of the α and β magnetization vectors.

If an experiment requires that the heteronuclear coupling *not* be refocused, then 180° pulses must be applied to both nuclei simultaneously; for the case of a ^{1}H nucleus coupled to a ^{13}C an additional 180° pulse must be applied to the ^{13}C spins at the same time as the proton 180° pulse.

We will return to echo modulation in Chapter 3; as we shall see, it is a crucial element in a large number of NMR experiments.

2 Fourier transform NMR

2.1 Introduction

This chapter reviews some of the basic elements of Fourier transform NMR spectroscopy, including a brief introduction to two-dimensional NMR. We shall not refer to specific experiments, but will attempt to lay the foundations for the more detailed discussions that follow. As with the previous chapter, the material should be at least vaguely familiar to anyone who uses, or is learning to use, a modern NMR spectrometer.

See also OCP 32, Sections 6.2–6.4.

2.2 Detection of the NMR signal

As described in Chapter 1, a pulse creates magnetization that precesses in the $x'y'$ plane of the laboratory frame. This magnetization is detected by the coil in the NMR probe and constitutes the *free induction decay*. All the frequencies detected, however, will be of the form $\omega_{rf} + \Omega$, where the transmitter frequency $\omega_{rf}/2\pi$ is (say) 600 MHz while the resonance offsets Ω are of the order of the chemical shift range of the nucleus being studied, typically a few kHz at most. The signal detected in the coil is 'mixed' with the basic spectrometer reference frequency (ω_{rf}) so that only the offset frequencies Ω are passed to the analogue-to-digital converter for digitization. The result of mixing the detected signal with ω_{rf} is precisely equivalent to the spectrometer observing the magnetization in a reference frame rotating at ω_{rf}. The rotating frame is therefore not just a convenient mathematical fiction; it actually coincides neatly with the experimental reality. The spectrometer reference frequency has a defined phase, so detection is along a fixed axis of the rotating frame.

ω_{rf}, like ω_0 and Ω, is an *angular* frequency (radians s^{-1}).

The free induction decay is the sum of many oscillating waves of differing frequencies, amplitudes, and phases. It is detected using two orthogonal (in the rotating frame) detection channels along (say) the x and y axes. This is known as *quadrature detection* and it is used by all modern NMR spectrometers. For each resonance in the spectrum, the two signals so acquired are cosine and sine functions of the offset frequency Ω, decaying at a rate $1/T_2$, as indicated in Fig. 2.1. They can be regarded as the real and imaginary parts of a complex time-domain signal $s(t)$ of the general form:

$$\begin{aligned} s(t) &= [\cos \Omega t + i \sin \Omega t] \exp(-t/T_2) \\ &= \exp(i\Omega t)\exp(-t/T_2) \qquad t \geq 0 \\ s(t) &= 0 \qquad t < 0 \end{aligned} \tag{2.1}$$

N.B. $i=\sqrt{-1}$ and $\exp(iA) = \cos A + i \sin A$

which can be converted into the frequency-domain function or *spectrum* $S(\omega)$ by Fourier transformation

$$S(\omega) = \int_{-\infty}^{\infty} s(t) \exp(-i\omega t)\, dt. \tag{2.2}$$

Thus,

$$S(\omega) = A(\Delta\omega) - iD(\Delta\omega) \tag{2.3}$$

with

$$A(\Delta\omega) = \frac{1/T_2}{(1/T_2)^2 + (\Delta\omega)^2} \quad \text{and} \quad D(\Delta\omega) = \frac{\Delta\omega}{(1/T_2)^2 + (\Delta\omega)^2}. \tag{2.4}$$

The frequency offset $\Delta\omega = \omega - \Omega$ is measured with respect to the centre of the resonance, Ω. The real part, $A(\Delta\omega)$, of the spectrum is an *absorptive* Lorentzian curve, centred on frequency Ω with a full width at half-height of $1/\pi T_2$ (measured in Hz) while the imaginary part, $D(\Delta\omega)$, is the corresponding *dispersive* Lorentzian (Fig. 2.1). In practice, only the real part of the spectrum is retained: absorptive lines are narrower than their dispersive counterparts and have their maximum amplitude at the frequency of interest.

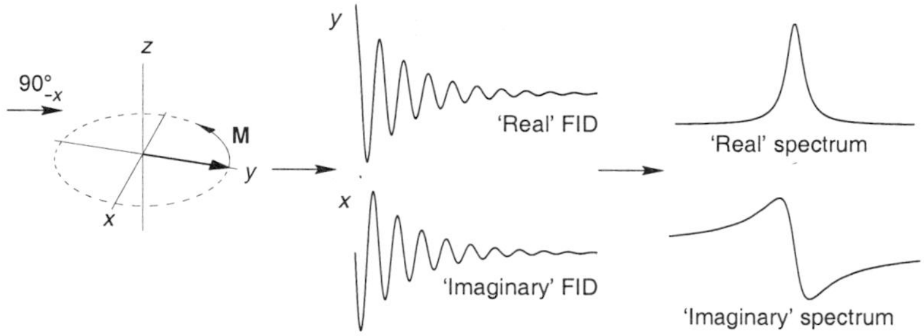

Fig. 2.1 Quadrature detection: simultaneous measurement of the x and y components of the free induction decay, using detectors fed with reference frequencies that differ in phase by 90°.

If detection is carried out along only one axis of the rotating frame, it is impossible to determine the sense of precession of the magnetization vectors, i.e. the signs of the frequencies in the NMR spectrum. This can be seen by replacing the $\exp(i\Omega t)$ term in Eqn 2.1 by $\cos\Omega t$, which equals $[\exp(+i\Omega t) + \exp(-i\Omega t)]/2$, or by $\sin\Omega t$, which equals $[\exp(+i\Omega t) - \exp(-i\Omega t)]/2i$, and Fourier transforming. Quadrature detection allows the spectrometer reference frequency, ω_{rf}, to be placed in the centre of the spectrum.

In reality, the NMR spectrum rarely initially turns out as in Fig. 2.1, with the absorption lineshape in the real part of the spectrum and the dispersion lineshape in the imaginary part. The two main reasons for this are (i) it is only by coincidence that the pulse aligns the initial magnetization with the real channel of the detector, and (ii) the pulse has finite length and the spectrometer always inserts a short delay between the end of the pulse and the start of acquisition; during both these times the various magnetization vectors develop phases that vary linearly with their resonance offset. So, immediately after Fourier transformation, a typical line in the real part of an NMR spectrum has arbitrary phase. Therefore, in order to obtain an absorption lineshape in the real part of the spectrum, one must 'phase the spectrum', which involves taking linear combinations of the real and imaginary parts until the desired result is achieved (Fig. 2.2).

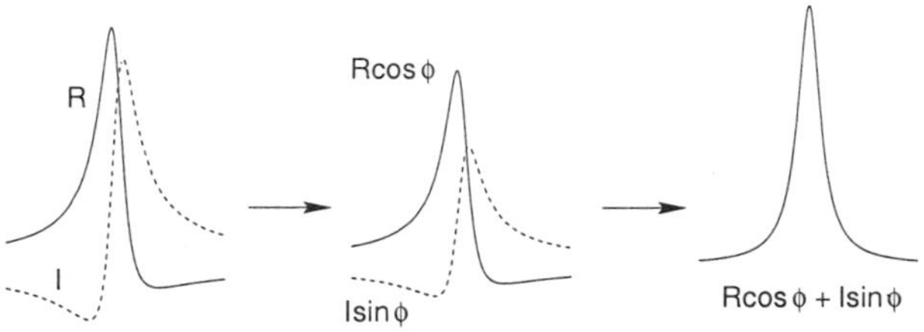

Fig. 2.2 Complex Fourier transformation of a free induction decay with quadrature detection yields a spectrum in need of phase correction. A linear combination of the real (R) and imaginary (I) parts of the spectrum gives absorption-phase lineshapes in the real part.

Although the spectrum $S(\omega)$ above was obtained analytically, this is not how the Fourier transform is done in the spectrometer computer. The free induction decay is first converted from analogue to digital form by sampling the oscillating signal at equal time intervals before the spectrum is calculated using an efficient numerical Fourier transform algorithm. The interval between samples, Δt, is determined by the so-called Nyquist condition, $\Delta t = 1/SW$, where SW is the desired spectral width, i.e. the spectrum extends $SW/2$ either side of the reference frequency. This corresponds to performing at least two samples per cycle of every oscillation present in the free induction decay.

2.3 Two-dimensional NMR

According to ancient tradition, the pulse sequence for a generic two-dimensional experiment is broken down into four sequential steps:

Preparation — Evolution — Mixing — Detection.

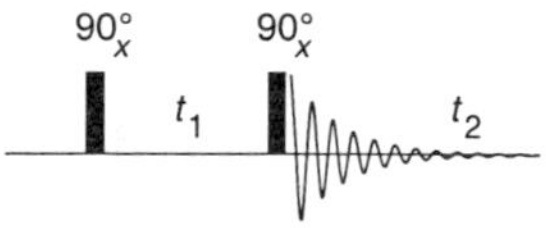

Fig. 2.3 A simple pulse sequence for two-dimensional NMR.

The free induction decay is acquired in the detection period, the time axis of which is normally labelled t_2. In most of the two-dimensional experiments that will concern us, the evolution period consists of uninterrupted free precession for a time t_1. Hence, one often talks of the t_1 *period* and the t_2 *period* of a two-dimensional experiment. The most basic form that the preparation period can take is a single 90°_x pulse to excite transverse magnetization. Similarly, one of the simplest mixing steps is also a 90°_x pulse. This pulse sequence is illustrated in Fig. 2.3.

The meaning of the term *mixing* will become apparent later.

Once again, imagine an NMR spectrum consisting of a single peak. During t_1 the magnetization vector precesses at a rate given by the offset Ω until the second 90°_x pulse flips the vector into the xz plane (Fig. 2.4). The component of **M** that is left along the z axis after the second 90° pulse does not contribute to the signal observed in t_2.

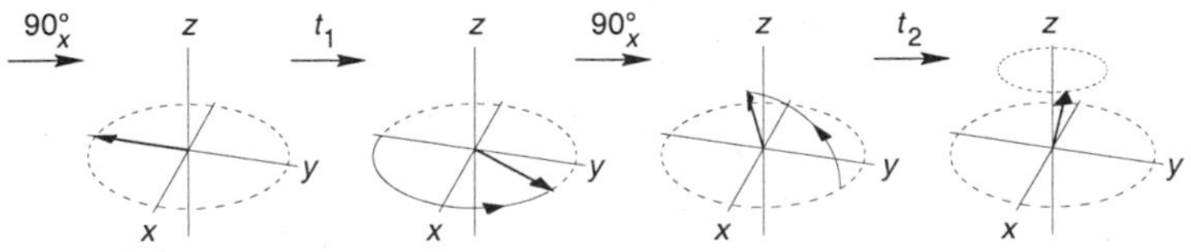

Fig. 2.4 Evolution of the magnetization vector during the pulse sequence of Fig. 2.3. The initial amplitude of the free induction decay at $t_2 = 0$ is determined by Ωt_1, the angle through which **M** precesses during t_1.

As indicated in Fig. 2.5a, the amplitude of the observed free induction decay (the x component of **M** after the pulse) will therefore depend on the duration of the evolution period t_1. Therefore, a series of experiments performed at increasing values of the delay t_1 gives rise to a *modulated* series of spectra after Fourier transformation, (Fig. 2.5b). If the column of data points labelled C in Fig. 2.5b, corresponding to the centre of the peak, is taken from the modulated data set then it will look exactly like a free induction decay except that its time axis is not t_2 but t_1, (Fig. 2.5c). This *interferogram* can be Fourier transformed in the usual way to give the normal NMR spectrum once again, (Fig. 2.5d). The Fourier transform of column B gives the same spectrum in this case, but with lower amplitude. By now it should be clear that a full two-dimensional spectrum can be arrived at by Fourier transforming *all* the columns of data points in (Fig. 2.5b). More formally, the result of performing a two-dimensional experiment is a time-domain data set $s(t_1, t_2)$; the free induction decays are all Fourier transformed to give a series of modulated spectra $S(t_1, F_2)$ and, finally, the interferograms are Fourier transformed to give a two-dimensional spectrum $S(F_1, F_2)$. When performing the two-dimensional experiment the magnetizations evolving during t_1 must be sampled regularly according to the same constraints as the data points in t_2. If a spectral width of SW Hertz is required (i.e. $SW/2$ either side of the transmitter frequency), the interval between t_1 (or t_2) samples should be $1/SW$ seconds, assuming quadrature detection is used.

The type of modulation described above is known as *amplitude modulation*. The free induction decays observed in t_2 are modulated in amplitude as a function of t_1 but their initial phases are always the same. The majority of important two-dimensional experiments (COSY, NOESY, etc.) inherently yield amplitude-modulated data, which has the advantage that the spectra are easily processed to give pure absorption two-dimensional lineshapes. The disadvantage is that the sense of the precession of the magnetization during t_1 is not determined (the same problem as experienced when using a single channel detector in one-dimensional NMR). Special methods have been developed to solve this problem (see below). Experiments that do not possess a mixing period generally yield *phase-modulated* data as a function of t_1. The best known examples are some NMR imaging experiments and the class of two-dimensional techniques known as *J spectroscopy*, neither of which will be discussed here.

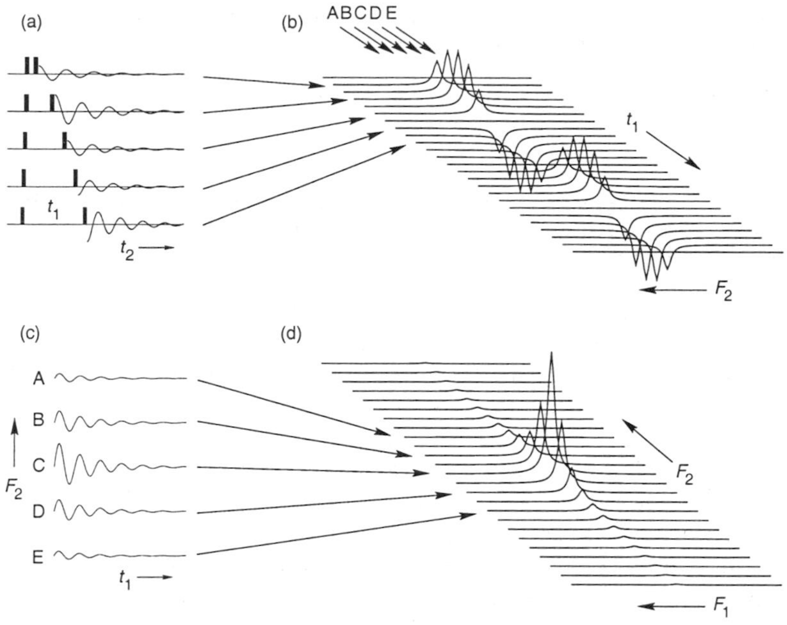

Fig. 2.5 The operation of a two-dimensional NMR experiment (Fig. 2.3) on a sample with a single NMR line. (a) Free induction decays for different values of the inter-pulse delay t_1. (b) The spectra obtained by Fourier transforming several such signals. (c) Interferograms constructed by extracting columns of data points from (b) at the indicated positions (A–E). (d) The two-dimensional spectrum that results from Fourier transformation of the interferograms. Only a few of the free induction decays and interferograms are shown for clarity.

2.4 Pure phase two-dimensional spectra

The simple two-dimensional experiment discussed in Section 2.3 gives data of the general form:

$$s(t_1,t_2) = \cos \Omega t_1 \exp(-t_1 / T_2)\exp(i\Omega t_2)\exp(-t_2 / T_2) \qquad (2.5)$$

where T_2 is the spin–spin relaxation time. In Section 2.2, we saw that the complex Fourier transform of a signal of the type $\exp(i\Omega t)\ \exp(-t/T_2)$ is $A(\Delta\omega) - iD(\Delta\omega)$ with $\Delta\omega = \omega - \Omega$. We will now abbreviate the notation further by writing $A_k^{\pm}$ and $D_k^{\pm}$ for absorptive and dispersive lineshapes centred at frequency $\pm\Omega$ in the F_k dimension. Complex Fourier transformation of $\exp(\pm i\Omega t_k)\ \exp(-t_k/T_2)$ thus gives $A_k^{\pm} - iD_k^{\pm}$. The Fourier transform of $s(t_1, t_2)$ with respect to t_2 is therefore

$$\begin{aligned} S(t_1,F_2) &= \cos \Omega t_1 \exp(-t_1 / T_2)\left[A_2^+ - iD_2^+\right] \\ &= \tfrac{1}{2}\left[\exp(+i\Omega t_1) + \exp(-i\Omega t_1)\right]\exp(-t_1 / T_2)\left[A_2^+ - iD_2^+\right]. \end{aligned} \qquad (2.6)$$

A complex Fourier transform of this with respect to t_1 gives

$$\begin{aligned} S(F_1,F_2) &= \tfrac{1}{2}\left[A_1^+ - iD_1^+ + A_1^- - iD_1^-\right]\left[A_2^+ - iD_2^+\right] \\ &= \tfrac{1}{2}\Big[(A_1^+A_2^+ - D_1^+D_2^+) + (A_1^-A_2^+ - D_1^-D_2^+) \\ &\quad - i(A_1^+D_2^+ + D_1^+A_2^+) - i(A_1^-D_2^+ + D_1^-A_2^+)\Big]. \end{aligned} \tag{2.7}$$

The mixture of two-dimensional absorption and dispersion lineshapes represented by terms such as $(A_1{}^+A_2{}^+ - D_1{}^+D_2{}^+)$ is the dreaded *phase twist* lineshape (Fig. 2.6). The real part of the two-dimensional spectrum described by Eqn 2.7 is thus a phase twist at a two-dimensional frequency $(F_1, F_2) = (+\Omega, +\Omega)$ and a phase twist at frequency $(F_1, F_2) = (-\Omega, +\Omega)$. This is the worst of both worlds. We have obtained phase twist lineshapes and have not achieved quadrature detection in F_1 and so cannot distinguish positive and negative frequencies in F_1. Clearly, we are doing something wrong.

The answer is to record and store separately two complete two-dimensional time-domain signals, one with cosine modulation as above and the other with sine modulation with respect to t_1. The latter is obtained by repeating the whole experiment with different pulse phases. Thus we have

$$\begin{aligned} s_{\cos}(t_1,t_2) &= \cos(\Omega t_1)\exp(-t_1/T_2)\exp(i\Omega t_2)\exp(-t_2/T_2) \\ s_{\sin}(t_1,t_2) &= \sin(\Omega t_1)\exp(-t_1/T_2)\exp(i\Omega t_2)\exp(-t_2/T_2). \end{aligned} \tag{2.8}$$

Fourier transformation of these two signals with respect to t_2 gives

$$\begin{aligned} S_{\cos}(t_1,F_2) &= \cos(\Omega t_1)\exp(-t_1/T_2)\left[A_2^+ - iD_2^+\right] \\ S_{\sin}(t_1,F_2) &= \sin(\Omega t_1)\exp(-t_1/T_2)\left[A_2^+ - iD_2^+\right]. \end{aligned} \tag{2.9}$$

The imaginary parts are deleted,

$$\begin{aligned} S_{\cos}(t_1,F_2) &= \cos(\Omega t_1)\exp(-t_1/T_2)\left[A_2^+\right] \\ S_{\sin}(t_1,F_2) &= \sin(\Omega t_1)\exp(-t_1/T_2)\left[A_2^+\right], \end{aligned} \tag{2.10}$$

and the latter is multiplied by i and added to the former,

$$S_{\cos}(t_1,F_2) + iS_{\sin}(t_1,F_2) = \exp(i\Omega t_1)\exp(-t_1/T_2)\left[A_2^+\right]. \tag{2.11}$$

Finally, Fourier transformation with respect to t_1 gives

$$S(F_1,F_2) = \left[A_1^+ - iD_1^+\right]\left[A_2^+\right] = A_1^+A_2^+ - iD_1^+A_2^+ . \tag{2.12}$$

As desired, the real part represents a two-dimensional absorption lineshape (see Fig. 2.6) at a frequency $(F_1, F_2) = (+\Omega, +\Omega)$.

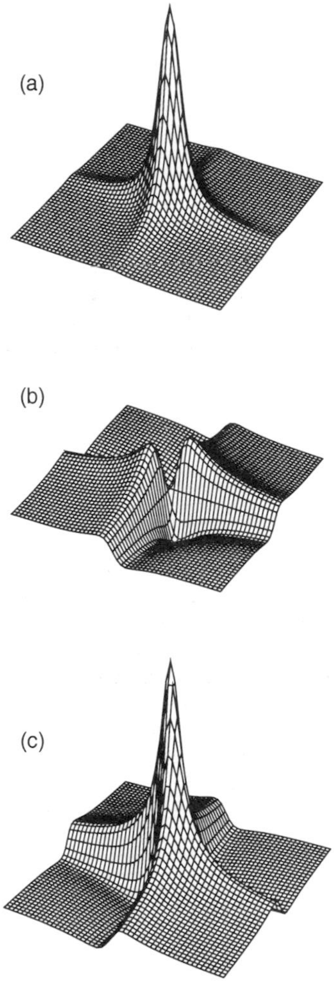

Fig. 2.6 Two-dimensional lineshapes: (a) absorption in both dimensions; (b) dispersion in both dimensions; (c) the phase twist lineshape, a combination of (a) and (b), whose broad wings and negative regions make it highly undesirable.

3 Product operators I

3.1 Introduction

Product operators have been around for at least 30 years, but only became popular as a way of describing NMR experiments after a review by Ernst's group (Sørensen *et al.* 1983). As the word 'operator' indicates, they are inherently quantum-mechanical in nature; however, they can be successfully used with no quantum-mechanical knowledge whatsoever. All that must be learned are two or three simple rules, although some familiarity with basic trigonometry is useful. The great beauty of product operators is that they largely retain the geometrical picture and intuitive 'feel' of the vector model, while offering an exact analysis (under conditions of weak J coupling) of the inner workings of modern NMR experiments.

A pair of spins is said to be *weakly coupled* when its spin–spin coupling constant J is small compared with the difference in resonance frequencies due to chemical shifts.

3.2 Product operators for one spin

Four operators suffice to describe NMR experiments on molecules containing an isolated $I = 1/2$ spin. They are:

$$\tfrac{1}{2}E \quad I_x \quad I_y \quad I_z. \tag{3.1}$$

The first is simply half the identity operator, and is included for purely formal reasons. It is the remaining three operators that interest us. They correspond to the x, y, and z magnetization of a single spin, in the rotating frame (Fig. 3.1).

We use product operators here not as operators as such but as a way of describing the state of the spin system, e.g. before and after a pulse or during a delay. ('Delay' is an NMR colloquialism for an interval of free precession.) As we shall see in Part B, operators are needed to *calculate* the efects of pulses and delays. Here we simply provide 'recipes' that enable one to do these transformations.

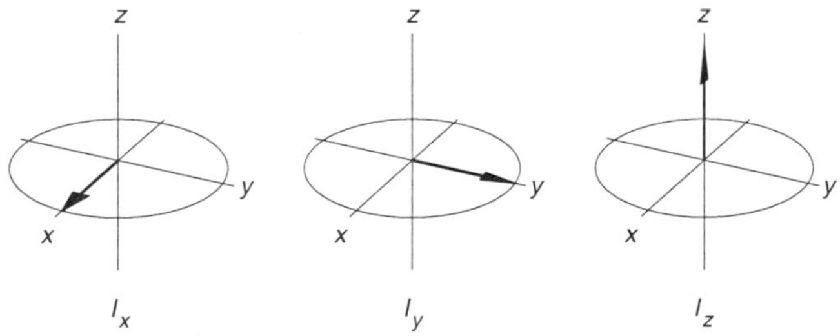

Fig. 3.1 Vector representation of the product operators I_x, I_y and I_z for a single spin-1/2 nucleus.

From the vector model, it is easy to see how I_x, I_y and I_z transform under (say) a 90° pulse.

$$\begin{array}{ll} I_x \xrightarrow{90^\circ_x} I_x & I_x \xrightarrow{90^\circ_y} -I_z \\ I_y \xrightarrow{90^\circ_x} I_z & I_y \xrightarrow{90^\circ_y} I_y \\ I_z \xrightarrow{90^\circ_x} -I_y & I_z \xrightarrow{90^\circ_y} I_x. \end{array} \quad (3.2)$$

More generally, the effect of pulses with flip angle β on the three operators is given by:

$$\begin{array}{ll} I_x \xrightarrow{\beta_x} I_x & I_x \xrightarrow{\beta_y} I_x \cos\beta - I_z \sin\beta \\ I_y \xrightarrow{\beta_x} I_y \cos\beta + I_z \sin\beta & I_y \xrightarrow{\beta_y} I_y \\ I_z \xrightarrow{\beta_x} I_z \cos\beta - I_y \sin\beta & I_z \xrightarrow{\beta_y} I_z \cos\beta + I_x \sin\beta. \end{array} \quad (3.3)$$

Similarly, it is easy to see from the vector model how the three operators transform as a result of free precession for a time t at a resonance offset Ω

$$\begin{array}{l} I_x \xrightarrow{\Omega t} I_x \cos\Omega t + I_y \sin\Omega t \\ I_y \xrightarrow{\Omega t} I_y \cos\Omega t - I_x \sin\Omega t \\ I_z \xrightarrow{\Omega t} I_z. \end{array} \quad (3.4)$$

As a reminder of the sign conventions involved in these rotations, which are the same as those we have used for the vector model, we can draw the hybrid vector model/product operator diagrams shown in Fig. 3.2.

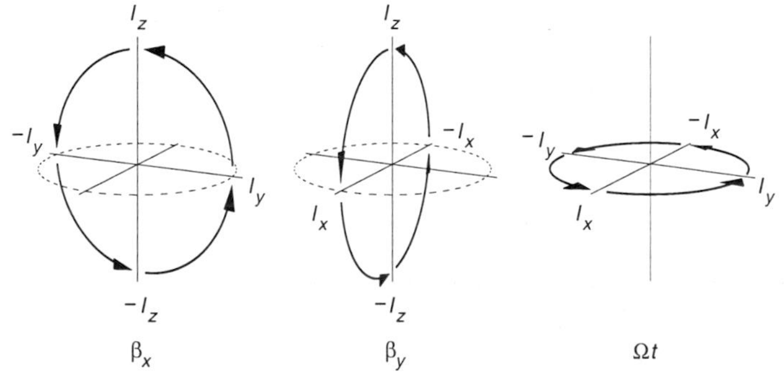

Fig. 3.2 The transformations of the product operators of a single spin-1/2 nucleus produced by x or y pulses (flip angle β), or a period t of free precession at a resonance offset Ω. These diagrams are nothing more than a pictorial representation of Eqns 3.3 and 3.4. For example, a β_y pulse transforms I_x into $I_x \cos\beta - I_z \sin\beta$, and $-I_z$ into $-I_z \cos\beta - I_x \sin\beta$.

Product operators can also be related to spectra and energy level diagrams, as indicated in Fig. 3.3 for a spin I = 1/2 nucleus. The dotted lines indicate the 90° phase difference between I_x and I_y operators. Thus, if we have two spins I and S, with the state of the system being described by $I_x + S_y$, then if the I peak is phased to pure absorption the S peak will be in dispersion (or vice versa).

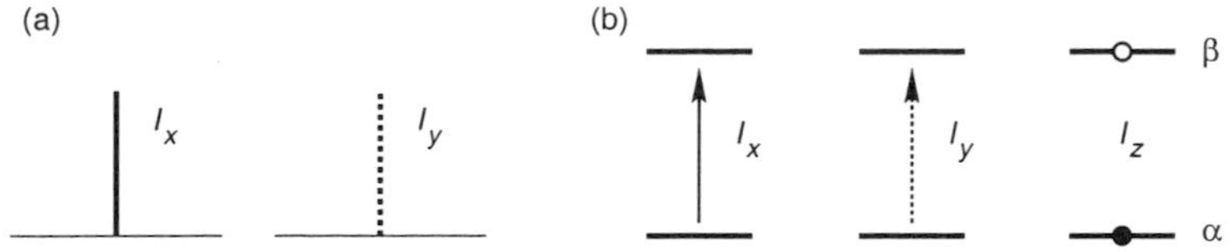

Fig. 3.3 Representation of the product operators for a single spin-1/2 nucleus in terms of (a) schematic spectra and (b) energy level diagrams. The dotted line for I_y is a reminder of the 90° phase difference from I_x. The arrows correspond to the transitions with which product operators are associated. The open and closed circles indicate, respectively, a population deficit and a population excess in the corresponding energy level, as would be found at thermal equilibrium, for example.

To illustrate the use of product operators, we can look at the effect of the spin echo sequence $90°_x - t - 180°_y - t'$ on a molecule with a one-line NMR spectrum. At each stage in this 'calculation' the rules in Eqns 3.3 and 3.4 are applied to each of the product operators in turn, starting with I_z:

$$\begin{aligned}
I_z &\xrightarrow{90^\circ_x} -I_y \\
&\xrightarrow{\Omega t} -I_y \cos\Omega t + I_x \sin\Omega t \\
&\xrightarrow{180^\circ_y} -I_y \cos\Omega t - I_x \sin\Omega t \\
&\xrightarrow{\Omega t'} -(I_y \cos\Omega t' - I_x \sin\Omega t')\cos\Omega t \\
&\qquad -(I_x \cos\Omega t' + I_y \sin\Omega t')\sin\Omega t \\
&= -I_y \cos\Omega t \cos\Omega t' + I_x \cos\Omega t \sin\Omega t' \\
&\qquad - I_x \sin\Omega t \cos\Omega t' - I_y \sin\Omega t \sin\Omega t'.
\end{aligned} \tag{3.5}$$

An echo is formed when the two free precession times either side of the 180° pulse, t and t', are equal. So, setting $t = t'$, it can be seen that the I_x terms cancel and we are left with:

$$-I_y \cos^2 \Omega t - I_y \sin^2 \Omega t = -I_y, \tag{3.6}$$

N.B. $\cos^2 A + \sin^2 A = 1$

i.e. the same state as was created by the initial $90°_x$ pulse.

However, all this may seem a bit of a cheat; although the product operators give the same result as the vector model, this is hardly surprising given that we have simply used them as a sort of 'vector model with numbers on it'. It is when we consider systems of J-coupled nuclei that product operators come into their own and we can start to explain experiments that cannot be understood using the vector model.

3.3 Product operators for two coupled spins

We will consider the NMR spectrum of two *weakly* J-coupled spin-1/2 nuclei, labelled I and S.

The restriction to weak coupling is the one serious limitation of the product operator approach (we deal with strong coupling in Part B).

The schematic spectrum in Fig. 3.4 may be taken to represent a homonuclear spin system, where I and S are both (say) protons, or a heteronuclear spin system (e.g. where I is ^{1}H and S is ^{13}C), in which case S is typically several-hundred megahertz off-resonance and is not affected by any pulses applied to the I spin. As the name implies, the 16 product operators for a system of two spin-1/2 nuclei can be formed by taking the products of the four operators for the individual spins I and S

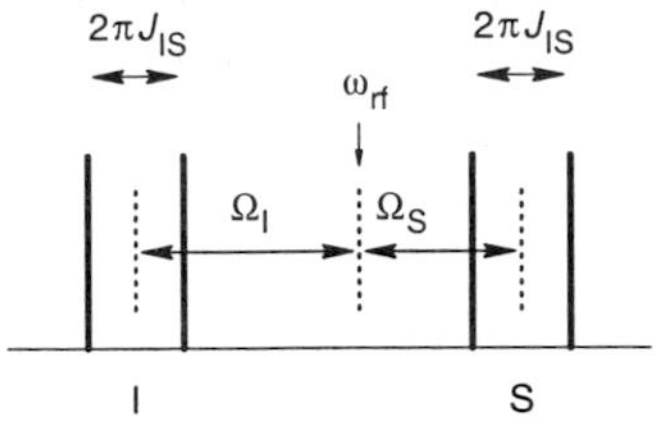

Fig. 3.4 Schematic NMR spectrum of a pair of spin-1/2 nuclei, I and S. ω_{rf} is the transmitter frequency, Ω_I and Ω_S are the offset frequencies and J_{IS} the spin–spin coupling constant. ω_{rf}, Ω_I, Ω_S and $2\pi J_{IS}$ are all angular frequencies (radians per second).

$$2\times\ \begin{array}{c|cccc} & \frac{1}{2}E & S_x & S_y & S_z \\ \hline \frac{1}{2}E & \frac{1}{2}E & S_x & S_y & S_z \\ I_x & I_x & 2I_xS_x & 2I_xS_y & 2I_xS_z \\ I_y & I_y & 2I_yS_x & 2I_yS_y & 2I_yS_z \\ I_z & I_z & 2I_zS_x & 2I_zS_y & 2I_zS_z \end{array} \qquad (3.7)$$

The factor of two on the left is one of those normalizations loved by theoreticians. For the moment we will ignore the four product operators in the centre of the table ($2I_xS_x$, $2I_xS_y$, $2I_yS_x$ and $2I_yS_y$). It will be shown later (Section 4.3) that these represent *multiple-quantum coherences*. The remaining product operators can be related to the vector model, as shown in Fig. 3.5. I_x, I_y, I_z, S_x, S_y and S_z are referred to as *in-phase* operators, while $2I_xS_z$, $2I_yS_z$, $2I_zS_x$, $2I_zS_y$ and $2I_zS_z$ are *antiphase* operators. Once again, the product operators can be related to schematic NMR spectra and energy level diagrams (Fig. 3.6).

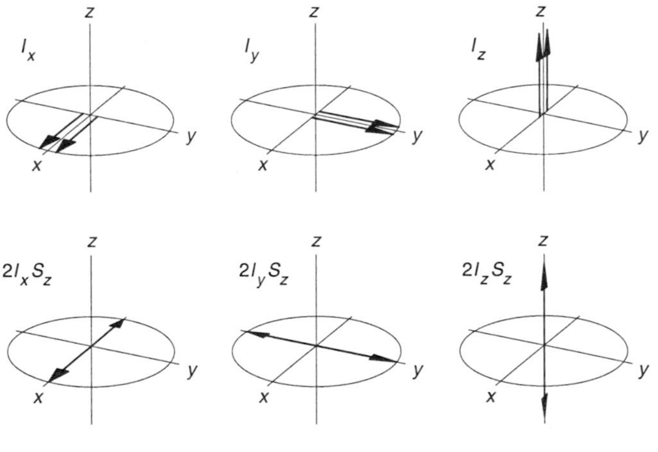

Fig. 3.5 Vector representation of some of the product operators for a two-spin IS system.

Finally, the *new* transformation rules, summarized in Fig. 3.7, for evolution under a J coupling J_{IS} for a period t are:

$$I_x \xrightarrow{\pi J_{IS}t} I_x \cos \pi J_{IS}t + 2I_yS_z \sin \pi J_{IS}t$$

$$I_y \xrightarrow{\pi J_{IS}t} I_y \cos \pi J_{IS}t - 2I_xS_z \sin \pi J_{IS}t$$

$$I_z \xrightarrow{\pi J_{IS}t} I_z$$

$$
\begin{aligned}
2I_xS_z &\xrightarrow{\pi J_{\mathrm{IS}}t} 2I_xS_z\cos\pi J_{\mathrm{IS}}t + I_y\sin\pi J_{\mathrm{IS}}t \\
2I_yS_z &\xrightarrow{\pi J_{\mathrm{IS}}t} 2I_yS_z\cos\pi J_{\mathrm{IS}}t - I_x\sin\pi J_{\mathrm{IS}}t \\
2I_zS_z &\xrightarrow{\pi J_{\mathrm{IS}}t} 2I_zS_z,
\end{aligned}
\tag{3.8}
$$

and similarly for S_x, S_y, $2I_zS_x$, and $2I_zS_y$.

We will see later where the πJ_{IS} frequency in these expressions comes from. For now, note that they are reassuringly consistent with the vector model (Figs 1.6 and 1.7). The $\pi J_{\mathrm{IS}}t$ label on the arrows is used simply as an abbreviation for 'evolution for a time t under the IS J coupling'.

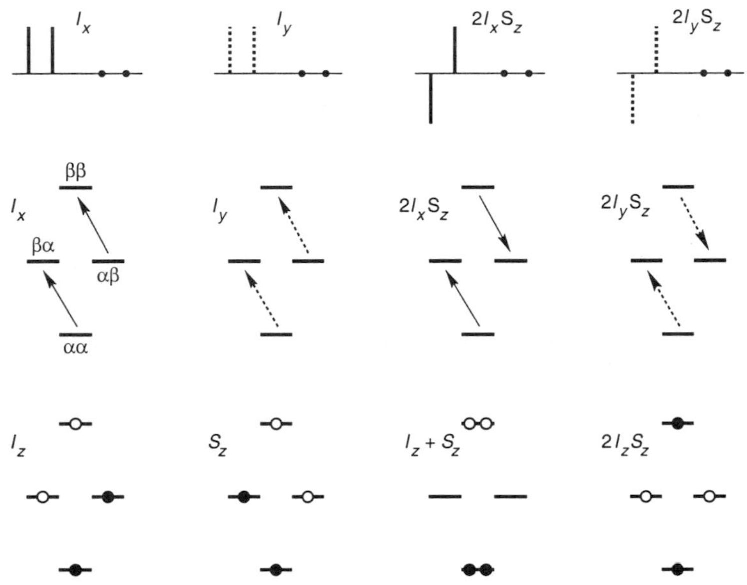

Fig. 3.6 Representation of some of the product operators for a two-spin IS system in terms of schematic spectra (above) and energy level diagrams (below). The corresponding S-spin operators have the same form, centred on Ω_{S}. Dotted lines, arrows and open/closed circles have the same significance as in Fig. 3.3.

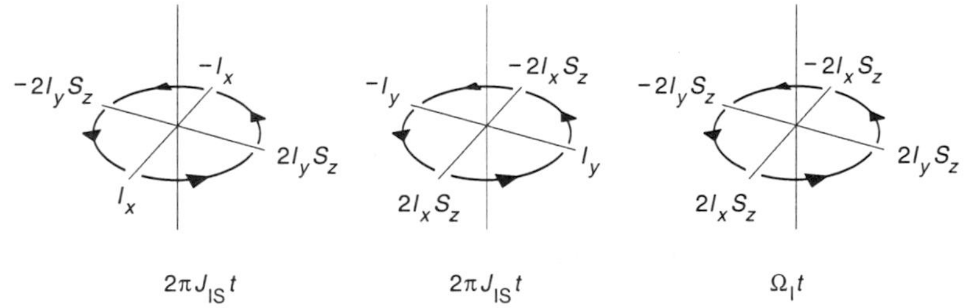

Fig. 3.7 The transformations of the product operators for a two-spin IS system produced by the scalar coupling J_{IS} and the resonance offset Ω_{I} during a period t.

The transformation rules for pulses and free precession under the resonance offset follow immediately from Eqns 3.3 and 3.4. For example:

$$
\begin{aligned}
2I_xS_z &\xrightarrow{\Omega_\mathrm{I}t} 2I_xS_z\cos\Omega_\mathrm{I}t + 2I_yS_z\sin\Omega_\mathrm{I}t \\
2I_zS_y &\xrightarrow{\Omega_\mathrm{S}t} 2I_zS_y\cos\Omega_\mathrm{S}t - 2I_zS_x\sin\Omega_\mathrm{S}t \\
2I_xS_z &\xrightarrow{(\beta_y)_\mathrm{I}} 2I_xS_z\cos\beta - 2I_yS_z\sin\beta \\
2I_zS_y &\xrightarrow{(90^\circ_x)_\mathrm{IS}} -2I_yS_z.
\end{aligned} \tag{3.9}
$$

All this may seem pretty abstract and of questionable value. So let us analyse in some detail a couple of simple examples where we know what the answer should be from the vector model.

3.4 Spin echoes

The product operator description of a spin echo in a J-coupled two-spin IS system starts as follows (the algebra is not as frightful as it looks). Concentrating on spin I, we have:

Although the evolutions caused by the chemical shift (Ω_I) and the spin–spin coupling (πJ_IS) actually occur simultaneously, we can treat them as if they were sequential; the order is immaterial. This will be justified in Chapters 8 and 9.

$$
\begin{aligned}
I_z &\xrightarrow{(90^\circ_x)_\mathrm{IS}} -I_y \\
&\xrightarrow{\Omega_\mathrm{I}t} -I_y\cos\Omega_\mathrm{I}t + I_x\sin\Omega_\mathrm{I}t \\
&\xrightarrow{\pi J_\mathrm{IS}t} -I_y\cos\Omega_\mathrm{I}t\cos\pi J_\mathrm{IS}t + 2I_xS_z\cos\Omega_\mathrm{I}t\sin\pi J_\mathrm{IS}t \\
&\qquad + I_x\sin\Omega_\mathrm{I}t\cos\pi J_\mathrm{IS}t + 2I_yS_z\sin\Omega_\mathrm{I}t\sin\pi J_\mathrm{IS}t.
\end{aligned} \tag{3.10}
$$

The 180°_y pulse in a homonuclear spin system changes the sign of all x and z operators of both spins:

Note that $2I_xS_z$ changes sign *twice* and so retains its original sign.

$$
\begin{aligned}
&\xrightarrow{(180^\circ_y)_\mathrm{IS}} -I_y\cos\Omega_\mathrm{I}t\cos\pi J_\mathrm{IS}t + 2I_xS_z\cos\Omega_\mathrm{I}t\sin\pi J_\mathrm{IS}t \\
&\qquad - I_x\sin\Omega_\mathrm{I}t\cos\pi J_\mathrm{IS}t - 2I_yS_z\sin\Omega_\mathrm{I}t\sin\pi J_\mathrm{IS}t.
\end{aligned} \tag{3.11}
$$

Now for the chemical shift evolution during the second free precession period:

$$
\begin{aligned}
&\xrightarrow{\Omega_\mathrm{I}t} -(I_y\cos\Omega_\mathrm{I}t - I_x\sin\Omega_\mathrm{I}t)\cos\Omega_\mathrm{I}t\cos\pi J_\mathrm{IS}t \\
&\qquad + (2I_xS_z\cos\Omega_\mathrm{I}t + 2I_yS_z\sin\Omega_\mathrm{I}t)\cos\Omega_\mathrm{I}t\sin\pi J_\mathrm{IS}t \\
&\qquad - (I_x\cos\Omega_\mathrm{I}t + I_y\sin\Omega_\mathrm{I}t)\sin\Omega_\mathrm{I}t\cos\pi J_\mathrm{IS}t \\
&\qquad - (2I_yS_z\cos\Omega_\mathrm{I}t - 2I_xS_z\sin\Omega_\mathrm{I}t)\sin\Omega_\mathrm{I}t\sin\pi J_\mathrm{IS}t.
\end{aligned} \tag{3.12}
$$

This looks horrible, but the terms in I_x cancel as do those containing $2I_yS_z$, to give

$$
\begin{aligned}
&- I_y\cos^2\Omega_\mathrm{I}t\cos\pi J_\mathrm{IS}t - I_y\sin^2\Omega_\mathrm{I}t\cos\pi J_\mathrm{IS}t \\
&+ 2I_xS_z\cos^2\Omega_\mathrm{I}t\sin\pi J_\mathrm{IS}t + 2I_xS_z\sin^2\Omega_\mathrm{I}t\sin\pi J_\mathrm{IS}t \\
= \; &- I_y\cos\pi J_\mathrm{IS}t + 2I_xS_z\sin\pi J_\mathrm{IS}t.
\end{aligned} \tag{3.13}
$$

Note that all terms involving Ωt have disappeared. This is as expected: the spin echo has refocused the chemical shift. To complete the calculation there must be evolution under the coupling in the second t period,

$$\begin{aligned} &\xrightarrow{\pi J_{IS}t} -I_y \cos^2 \pi J_{IS}t + 2I_xS_z \cos\pi J_{IS}t \sin\pi J_{IS}t \\ &\quad + 2I_xS_z \sin\pi J_{IS}t \cos\pi J_{IS}t + I_y \sin^2 \pi J_{IS}t \\ &= -I_y \cos 2\pi J_{IS}t + 2I_xS_z \sin 2\pi J_{IS}t. \end{aligned} \tag{3.14}$$

N.B. $\cos^2 A - \sin^2 A = \cos 2A$ and $2\sin A \cos A = \sin 2A$

Reassuringly, this is exactly the result predicted by the vector model. Only the J coupling has evolved, at a frequency πJ_{IS} for a time $2t$. Remembering that in this homonuclear spin system I behaves exactly the same as S, the full result becomes:

$$-(I_y + S_y)\cos 2\pi J_{IS}t + (2I_xS_z + 2I_zS_x)\sin 2\pi J_{IS}t. \tag{3.15}$$

In a *heteronuclear* IS spin system the 90° and 180° pulses only affect the I spin and we can write:

$$\begin{aligned} I_z &\xrightarrow{(90^\circ_x)_I} \xrightarrow{\Omega_I t} \xrightarrow{\pi J_{IS}t} \\ &\quad - I_y \cos\Omega_I t \cos\pi J_{IS}t + 2I_xS_z \cos\Omega_I t \sin\pi J_{IS}t \\ &\quad + I_x \sin\Omega_I t \cos\pi J_{IS}t + 2I_yS_z \sin\Omega_I t \sin\pi J_{IS}t \\ &\xrightarrow{(180^\circ_y)_I} -I_y \cos\Omega_I t \cos\pi J_{IS}t - 2I_xS_z \cos\Omega_I t \sin\pi J_{IS}t \\ &\quad - I_x \sin\Omega_I t \cos\pi J_{IS}t + 2I_yS_z \sin\Omega_I t \sin\pi J_{IS}t \\ &\xrightarrow{\Omega_I t} -I_y \cos\pi J_{IS}t - 2I_xS_z \sin\pi J_{IS}t. \end{aligned} \tag{3.16}$$

This is of course exactly the same as Eqn 3.10.

Note that only I_x is inverted by the $(180^\circ{}_y)_I$ pulse.

In this last step we have implicitly carried out the cancellations and trigonometric reduction which reveal that the chemical shift has been refocused. This expression differs from the homonuclear case in Eqn 3.13 only in the sign of the antiphase term $2I_xS_z$. Finally, the evolution under the coupling in the second t period:

$$\begin{aligned} &\xrightarrow{\pi J_{IS}t} -I_y \cos^2 \pi J_{IS}t + 2I_xS_z \cos\pi J_{IS}t \sin\pi J_{IS}t \\ &\quad - 2I_xS_z \sin\pi J_{IS}t \cos\pi J_{IS}t - I_y \sin^2 \pi J_{IS}t \\ &= -I_y. \end{aligned} \tag{3.17}$$

As predicted by the vector model, both the J coupling and the chemical shift have been refocused for the heteronuclear case.

4 Product operators II

4.1 Introduction

We now start to use product operators to describe pulse sequences that cannot be satisfactorily understood using the vector model. In this chapter, we deal with one-dimensional experiments.

4.2 INEPT

INEPT: insensitive nuclei enhanced by polarization transfer. NMR abounds with acronyms: not all are as ironic, whimsical or even as tasteful as this one.

γ, the gyromagnetic ratio of a nucleus determines the energy level splitting produced by a magnetic field and hence the population difference at equilibrium. See OCP 32, Table 1.3.

The INEPT experiment (Fig. 4.1) uses the magnetization of high γ nuclei (such as ^{1}H, ^{19}F, ^{31}P, etc.) to enhance the weak NMR signals of low γ nuclei (such as ^{13}C, ^{15}N, ^{57}Fe, etc.). The INEPT pulse sequence is also found as an element in many two-dimensional experiments such as the old-fashioned heteronuclear shift correlation and more modern 'inverse' techniques which have ^{13}C or ^{15}N single-quantum coherence evolving during t_1.

The free precession interval τ is set as near as possible to $1/4J_{\mathrm{IS}}$ where J_{IS} is the coupling between the high γ spin I and the low γ spin S. For example, the one-bond ^{13}C–^{1}H J coupling is typically about 150 Hz, giving $\tau = 1.67$ ms. The initial state of the IS spin system can be written $aI_z + bS_z$ where the coefficients a and b are there to remind us that in a heteronuclear spin system the z magnetizations associated with spins I and S have inherently different strengths. For example, for ^{1}H and ^{13}C, $a \approx 4b$ (because in this case $\gamma_{\mathrm{I}} \approx 4\gamma_{\mathrm{S}}$). The experiment starts with a $90°_x$ pulse on the I spin:

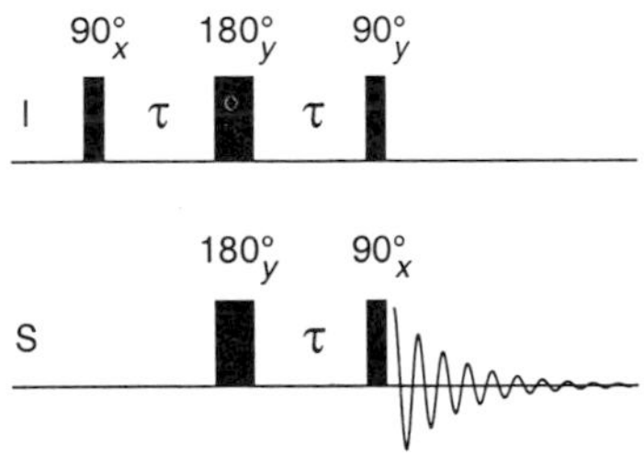

Fig. 4.1 The INEPT pulse sequence.

$$aI_z + bS_z \xrightarrow{(90^\circ_x)_{\mathrm{I}}} -aI_y + bS_z. \tag{4.1}$$

We now use the first of many short-cuts that can be introduced to make product operator calculations less tedious. The INEPT sequence contains a spin echo: therefore, to calculate the state of the spin system at the end of the second τ period we can assume that there is no resonance offset, Ω_{I}, and just have the system evolving under the J coupling J_{IS}. The order in which the two 180° pulses are treated is irrelevant.

$$\xrightarrow{\pi J_{\mathrm{IS}}\tau} -aI_y \cos\pi J_{\mathrm{IS}}\tau + a2I_xS_z \sin\pi J_{\mathrm{IS}}\tau + bS_z$$

$$\xrightarrow{(180^\circ_y)_{\mathrm{I}}} -aI_y \cos\pi J_{\mathrm{IS}}\tau - a2I_xS_z \sin\pi J_{\mathrm{IS}}\tau + bS_z$$

$$\xrightarrow{(180^\circ_y)_{\mathrm{S}}} -aI_y \cos\pi J_{\mathrm{IS}}\tau + a2I_xS_z \sin\pi J_{\mathrm{IS}}\tau - bS_z$$

$$\xrightarrow{\pi J_{IS}\tau} -aI_y \cos^2 \pi J_{IS}\tau + a2I_xS_z \cos\pi J_{IS}\tau \sin\pi J_{IS}\tau$$

$$+ a2I_xS_z \sin\pi J_{IS}\tau \cos\pi J_{IS}\tau + aI_y \sin^2 \pi J_{IS}\tau - bS_z \quad (4.2)$$

$$= -aI_y \cos 2\pi J_{IS}\tau + a2I_xS_z \sin 2\pi J_{IS}\tau - bS_z.$$

Now, if we substitute in the value $\tau = 1/4J_{IS}$ we find that the purpose of the τ intervals is to allow the spin I doublet to evolve into a purely *antiphase* state, which the 90° pulses transfer from I to S:

$$= a2I_xS_z - bS_z \xrightarrow{(90^\circ_y)_I} -a2I_zS_z - bS_z$$
$$\xrightarrow{(90^\circ_x)_S} a2I_zS_y + bS_y. \quad (4.3)$$

Both of these terms correspond to S-spin signals. The second represents in-phase y magnetization and has the normal (low) intensity. The first is antiphase y magnetization of spin S, but its intensity factor is that of spin I. Thus, if I is ^1H and S is ^{13}C, the first term has approximately four times the intensity of the second, while if I is ^1H and S is ^{15}N, a is approximately 10 times b. As shown in Fig. 4.2, we can think of each of these terms contributing towards the final S-spin spectrum.

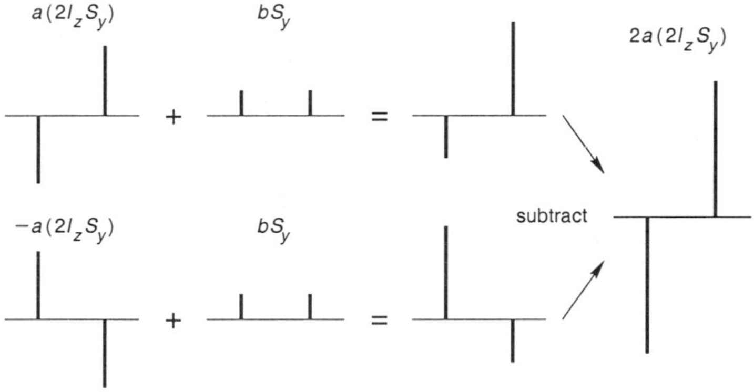

Fig. 4.2 INEPT phase cycling to cancel the bS_y contribution to the S-spin spectrum.

The asymmetry in the S spectrum can be removed by *phase cycling*, which is used to suppress the unenhanced spin S magnetization (the bS_y term). It is easily seen that if a second experiment is performed with the $90°_y$ pulse on the I spin replaced with a $90°_{-y}$ pulse then the final state of the system is $-a2I_zS_y + bS_y$. Thus, if this second spectrum is subtracted from the first rather than added to it, the final result is

$$(a2I_zS_y + bS_y) - (-a2I_zS_y + bS_y) = (2a)2I_zS_y. \quad (4.4)$$

This represents a purely antiphase S spectrum enhanced by a factor $a/b = \gamma_I/\gamma_S$. Nuclei with high γ usually relax faster than those with low γ and, since the repetition time of the INEPT sequence is determined by the relaxation of I magnetization, this represents an additional sensitivity advantage.

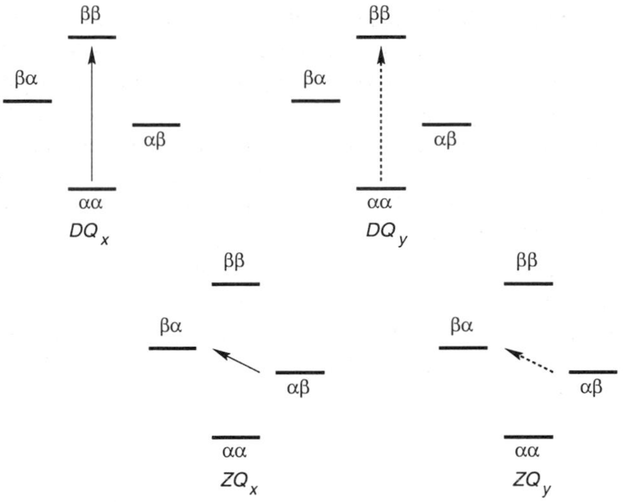

Fig. 4.3 Representation of the two-spin double- and zero-quantum product operators in terms of the energy level diagram. The dotted lines are a reminder of the 90° phase difference between x and y operators.

4.3 Multiple-quantum coherence

The term 'coherence' is used to indicate the precessing states that involve at least one x or y operator, and includes I_x, I_y, S_x and S_y, which represent single-quantum coherences. See also Chapters 6 and 8.

The two-spin product operators, $2I_xS_x$, $2I_yS_x$, $2I_xS_y$ and $2I_yS_y$, represent *multiple-quantum coherence*. These coherences are not directly observable in an NMR experiment since they induce no current in the receiver coil. They cannot be described using the vector model. From inspection of the energy level diagram for a two-spin system (Fig. 4.3) we would expect four multiple-quantum coherences. The transitions DQ_x and DQ_y are the two double-quantum coherences (90° out of phase with respect to one another); ZQ_x and ZQ_y are the two zero-quantum coherences. Unfortunately, all four are linear combinations of two-spin product operators:

$$DQ_x = \tfrac{1}{2}(2I_xS_x - 2I_yS_y) \qquad DQ_y = \tfrac{1}{2}(2I_yS_x + 2I_xS_y)$$
$$ZQ_x = \tfrac{1}{2}(2I_xS_x + 2I_yS_y) \qquad ZQ_y = \tfrac{1}{2}(2I_yS_x - 2I_xS_y). \tag{4.5}$$

Thus, for example, a two-spin operator $2I_xS_x$ corresponds to both double- and zero-quantum coherences: $2I_xS_x = DQ_x + ZQ_x$. Using Eqn 3.4 for both I and S, the evolution of these operators is fairly easily seen to be:

$$DQ_x \xrightarrow{(\Omega_\mathrm{I}+\Omega_\mathrm{S})t} DQ_x\cos(\Omega_\mathrm{I}+\Omega_\mathrm{S})t + DQ_y\sin(\Omega_\mathrm{I}+\Omega_\mathrm{S})t$$
$$ZQ_x \xrightarrow{(\Omega_\mathrm{I}+\Omega_\mathrm{S})t} ZQ_x\cos(\Omega_\mathrm{I}-\Omega_\mathrm{S})t + ZQ_y\sin(\Omega_\mathrm{I}-\Omega_\mathrm{S})t \tag{4.6}$$

and similarly for DQ_y and ZQ_y. Less obviously, IS zero- and double-quantum coherences do not evolve under the influence of the coupling J_IS. We will demonstrate this later (Section 9.7).

Perhaps the simplest NMR experiment to make use of multiple-quantum coherence is a one-dimensional *double-quantum filter* (Fig. 4.4). The symbol ϕ outside the brackets around the first three pulses indicates that a phase shift will be applied to these pulses on successive acquisitions as part of a phase cycle. Double-quantum coherence can only be excited in multi-spin systems and not in isolated spin-1/2 nuclei. Thus, a double-quantum filter can be used in ^{1}H spectroscopy to suppress solvent resonances (e.g. H_2O) and, under the acronym INADEQUATE, in ^{13}C spectroscopy, to suppress the signal from the 1% of molecules that contain a single ^{13}C nucleus in order to observe the 0.01% that contain two J-coupled ^{13}C spins. If we imagine a three-spin system consisting of two J-coupled spins I and S and a third, isolated, spin R whose signal we wish to suppress, then at the end of the second τ delay the state of the system is given by:

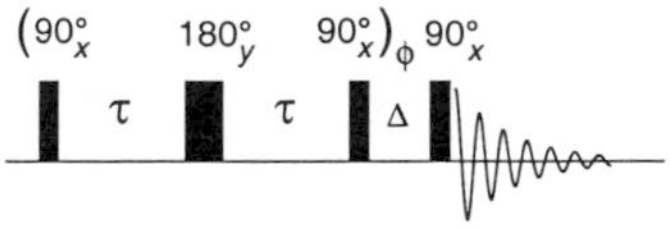

Fig. 4.4 Pulse sequence for the double-quantum filter (or INADEQUATE) experiment.

INADEQUATE: incredible natural abundance double-quantum transfer experiment.

$$-(I_y + S_y)\cos 2\pi J_{IS}\tau + (2I_xS_z + 2I_zS_x)\sin 2\pi J_{IS}\tau - R_y. \qquad (4.7)$$

Apart from the last term, this is just Eqn 3.15.

As in INEPT, the delay τ is chosen to be $1/4J_{IS}$ to maximize the antiphase component:

$$(2I_xS_z + 2I_zS_x) - R_y. \qquad (4.8)$$

The second 90° pulse now creates a state of pure IS double-quantum coherence (DQ_y)

$$\xrightarrow{(90^\circ_x)_{ISR}} -(2I_xS_y + 2I_yS_x) - R_z. \qquad (4.9)$$

The interval, Δ, is a very short delay (typically a few microseconds) to allow the spectrometer to change the pulse phases; any evolution during Δ can safely be neglected. The final 90° pulse then reconverts the IS double-quantum coherence back into observable antiphase magnetization:

$$\xrightarrow{(90^\circ_x)_{ISR}} -(2I_xS_z + 2I_zS_x) + R_y. \qquad (4.10)$$

However, no discrimination between the coupled spins I and S and the isolated spin R has yet been achieved. We must perform a phase cycle that selects double-quantum coherence during the delay, Δ, and rejects ordinary magnetization (single-quantum coherence). This is done by successively adding $\phi = 90°$ to the phase of the first three pulses over four experiments:

	Experiment	*Final state of system*
(1)	$90^\circ_{+x} - \tau - 180^\circ_{+y} - \tau - 90^\circ_{+x}\ 90^\circ_x$	$-(2I_xS_z + 2I_zS_x) + R_y$
(2)	$90^\circ_{+y} - \tau - 180^\circ_{-x} - \tau - 90^\circ_{+y}\ 90^\circ_x$	$+(2I_xS_z + 2I_zS_x) + R_y$ (4.11)
(3)	$90^\circ_{-x} - \tau - 180^\circ_{-y} - \tau - 90^\circ_{-x}\ 90^\circ_x$	$-(2I_xS_z + 2I_zS_x) + R_y$
(4)	$90^\circ_{-y} - \tau - 180^\circ_{+x} - \tau - 90^\circ_{-y}\ 90^\circ_x$	$+(2I_xS_z + 2I_zS_x) + R_y.$

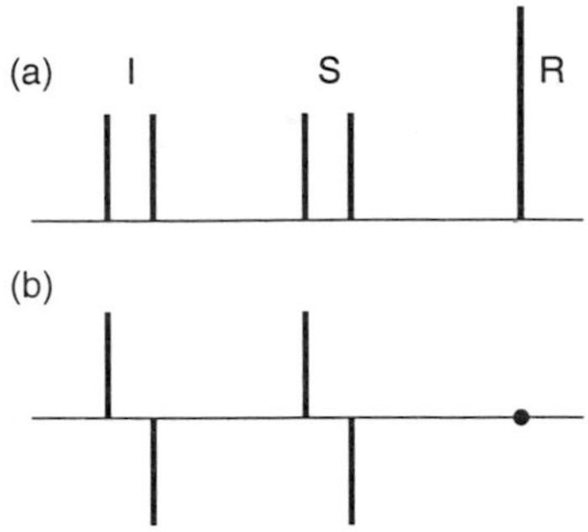

Fig. 4.5 Schematic spectra of an ISR spin system (a) following a 90° pulse and (b) after a double-quantum filter has been used to suppress the resonance of the uncoupled R spin.

Therefore if, instead of merely adding the free induction decays produced by these four experiments, one takes the combination (1) – (2) + (3) – (4), the final result is:

$$(1)-(2)+(3)-(4) = -4(2I_xS_z + 2I_zS_x)\,. \tag{4.12}$$

In this way, the resonance of spin R does not appear in the final spectrum (Fig. 4.5). A much more general way of describing phase cycling is presented in Chapter 6.

4.4 Multi-spin systems

Systems of three J-coupled spin-1/2 nuclei I, S and R are sufficiently complicated to show most of the various multiplet and flip angle effects encountered in two-dimensional NMR. There is rarely any need to do product operator calculations for four- and five-spin systems; the result can usually be predicted from knowledge of the three-spin case. There are 64 product operators for the ISR system and the reader should, if necessary, be able to write all of them down, deriving them from the two-spin operators in the same way as these were obtained from the one-spin operators. The ISR spin system is described by three resonance offsets, Ω_{I}, Ω_{S} and Ω_{R}, and three J couplings, J_{IS}, J_{IR} and J_{RS}. The new feature in the product operator description of an ISR system is the three-spin operators involving I, S and R. These appear as a result of (say) I_x magnetization evolving under the influence of both J_{IS} and J_{IR}:

$$\begin{aligned} I_x &\xrightarrow{\pi J_{\text{IS}}t} I_x \cos \pi J_{\text{IS}}t + 2I_yS_z \sin \pi J_{\text{IS}}t \\ &\xrightarrow{\pi J_{\text{IR}}t} I_x \cos \pi J_{\text{IS}}t \cos \pi J_{\text{IR}}t + 2I_yR_z \cos \pi J_{\text{IS}}t \sin \pi J_{\text{IR}}t \\ &\quad + 2I_yS_z \sin \pi J_{\text{IS}}t \cos \pi J_{\text{IR}}t - 4I_xS_zR_z \sin \pi J_{\text{IS}}t \sin \pi J_{\text{IR}}t. \end{aligned} \tag{4.13}$$

It is not always particularly useful to relate the three-spin product operators to the vector model or to energy level diagrams, which are a little too complicated to be an intuitive aid. We should now be sufficiently familiar with the product operators to accept a description in words:

$4I_xS_zR_z$ — x magnetization of spin I, antiphase with respect to spins S and R;

$4I_xS_xR_z$ — mixture of IS double- and zero-quantum coherences, both antiphase with respect to spin R;

$4I_xS_xR_x$ — mixture of ISR triple-quantum coherence and unobservable ISR three-spin single-quantum coherences;

$4I_zS_zR_z$ — a highly ordered population state.

It is useful, however, to relate the single-quantum product operators that occur in an ISR spin system to schematic spectra (Fig. 4.6). Any J coupling between S and R is irrelevant to these I-spin multiplets.

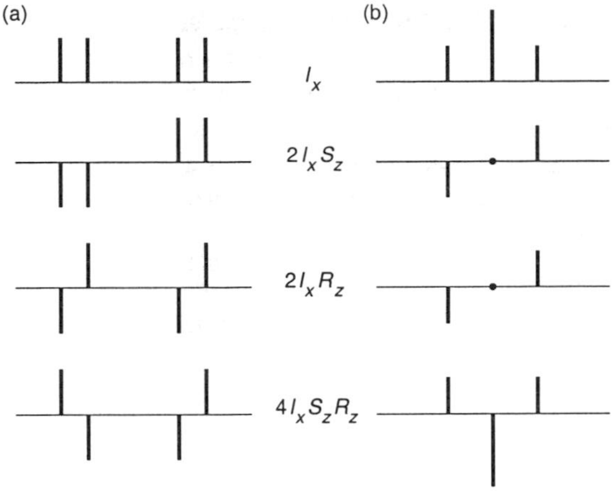

Fig. 4.6 Representation of the single-quantum I-spin product operators in an ISR spin system with (a) $J_{IS} > J_{IR}$ and (b) $J_{IS} = J_{IR}$.

4.5 DEPT

The DEPT experiment (Fig. 4.7) is a good example with which to end this chapter as it is a popular and widely-used NMR experiment, and its product operator description, involving as it does multiple-quantum coherence in multi-spin systems, draws together many of the themes introduced above. DEPT is used to 'edit' ^{13}C NMR spectra into subspectra containing only CH, CH_2 or CH_3 groups. Organic chemists find this useful.

DEPT: distortionless enhancement by polarization transfer

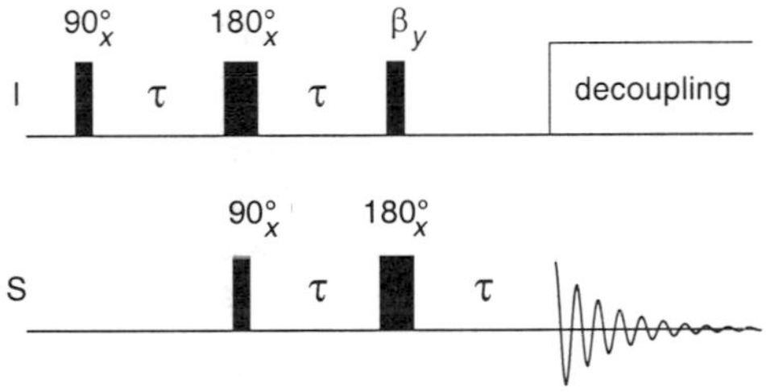

Fig. 4.7 The DEPT pulse sequence.

We can start by following the magnetization of just one proton (I_1), calling the others (if present) I_2 and I_3, all coupled to a ^{13}C nucleus S:

$$I_{1z} \xrightarrow{(90^\circ_x)_I} \xrightarrow{\Omega_I \tau} \xrightarrow{\pi J_{I_1S}\tau}$$

$$- I_{1y} \cos\Omega_I\tau \cos\pi J_{I_1S}\tau + 2I_{1x}S_z \cos\Omega_I\tau \sin\pi J_{I_1S}\tau \quad (4.14)$$
$$+ I_{1x} \sin\Omega_I\tau \cos\pi J_{I_1S}\tau + 2I_{1y}S_z \sin\Omega_I\tau \sin\pi J_{I_1S}\tau.$$

This is just Eqn 3.10.

The τ period is set to $1/2J_{I_1S}$ so that we get:

$$2I_{1x}S_z \cos\Omega_I\tau + 2I_{1y}S_z \sin\Omega_I\tau. \quad (4.15)$$

The next two pulses create a state of heteronuclear multiple-quantum (ZQ and DQ) coherence:

$$\xrightarrow{(180^\circ_x)_\mathrm{I}} 2I_{1x}S_z\cos\Omega_\mathrm{I}\tau - 2I_{1y}S_z\sin\Omega_\mathrm{I}\tau$$
$$\xrightarrow{(90^\circ_x)_\mathrm{S}} -2I_{1x}S_y\cos\Omega_\mathrm{I}\tau + 2I_{1y}S_y\sin\Omega_\mathrm{I}\tau. \qquad (4.16)$$

This multiple-quantum coherence evolves during the next free precession interval under the influence of both I and S chemical shifts:

$$\begin{aligned}\xrightarrow{\Omega_\mathrm{I}\tau} & -2I_{1x}S_y\cos\Omega_\mathrm{I}\tau\cos\Omega_\mathrm{I}\tau - 2I_{1y}S_y\cos\Omega_\mathrm{I}\tau\sin\Omega_\mathrm{I}\tau \\ & + 2I_{1y}S_y\sin\Omega_\mathrm{I}\tau\cos\Omega_\mathrm{I}\tau - 2I_{1x}S_y\sin\Omega_\mathrm{I}\tau\sin\Omega_\mathrm{I}\tau \\ & = -2I_{1x}S_y \\ \xrightarrow{\Omega_\mathrm{S}\tau} & -2I_{1x}S_y\cos\Omega_\mathrm{S}\tau + 2I_{1x}S_x\sin\Omega_\mathrm{S}\tau.\end{aligned} \qquad (4.17)$$

The I_1-spin chemical shift has been refocused, but not the S-spin shift (yet). These coherences also evolve under the influence of the $J_{\mathrm{I}_2\mathrm{S}}$ and $J_{\mathrm{I}_3\mathrm{S}}$ couplings, if present. As noted in Section 4.3, the $J_{\mathrm{I}_1\mathrm{S}}$ coupling does not cause the multiple-quantum coherences between I_1 and S spins to evolve. Therefore, the result of the evolution under J coupling during the second τ period depends on whether we are describing a CH, CH_2 or CH_3 group:

I_1, I_2, and I_3 are *equivalent*, so that $J_{\mathrm{I}_1\mathrm{S}} = J_{\mathrm{I}_2\mathrm{S}} = J_{\mathrm{I}_3\mathrm{S}} = J_{\mathrm{IS}}$.

$$\begin{aligned}&\mathrm{CH} \longrightarrow -2I_{1x}S_y\cos\Omega_\mathrm{S}\tau + 2I_{1x}S_x\sin\Omega_\mathrm{S}\tau, \\ &\mathrm{CH_2} \xrightarrow{\pi J_{\mathrm{I}_2\mathrm{S}}\tau\ (\tau=1/2J_\mathrm{IS})} 4I_{1x}I_{2z}S_x\cos\Omega_\mathrm{S}\tau + 4I_{1x}I_{2z}S_y\sin\Omega_\mathrm{S}\tau, \\ &\mathrm{CH_3} \xrightarrow{\pi J_{\mathrm{I}_2\mathrm{S}}\tau\ (\tau=1/2J_\mathrm{IS})} 4I_{1x}I_{2z}S_x\cos\Omega_\mathrm{S}\tau + 4I_{1x}I_{2z}S_y\sin\Omega_\mathrm{S}\tau \\ &\qquad \xrightarrow{\pi J_{\mathrm{I}_3\mathrm{S}}\tau\ (\tau=1/2J_\mathrm{IS})} 8I_{1x}I_{2z}I_{3z}S_y\cos\Omega_\mathrm{S}\tau - 8I_{1x}I_{2z}I_{3z}S_x\sin\Omega_\mathrm{S}\tau.\end{aligned} \qquad (4.18)$$

The next two pulses reconvert these heteronuclear multiple-quantum coherences into observable single-quantum coherences (non-observable terms are not shown below):

$$\begin{aligned}&\mathrm{CH} \xrightarrow{(\beta_y)_\mathrm{I}}\xrightarrow{(180^\circ_x)_\mathrm{S}} -2I_{1z}S_y\cos\Omega_\mathrm{S}\tau\sin\beta \\ &\qquad\qquad - 2I_{1z}S_x\sin\Omega_\mathrm{S}\tau\sin\beta, \\ &\mathrm{CH_2} \xrightarrow{(\beta_y)_\mathrm{I}}\xrightarrow{(180^\circ_x)_\mathrm{S}} -4I_{1z}I_{2z}S_x\cos\Omega_\mathrm{S}\tau\sin\beta\cos\beta \\ &\qquad\qquad + 4I_{1z}I_{2z}S_y\sin\Omega_\mathrm{S}\tau\sin\beta\cos\beta, \\ &\mathrm{CH_3} \xrightarrow{(\beta_y)_\mathrm{I}}\xrightarrow{(180^\circ_x)_\mathrm{S}} 8I_{1z}I_{2z}I_{3z}S_y\cos\Omega_\mathrm{S}\tau\sin\beta\cos^2\beta \\ &\qquad\qquad + 8I_{1z}I_{2z}I_{3z}S_x\sin\Omega_\mathrm{S}\tau\sin\beta\cos^2\beta.\end{aligned} \qquad (4.19)$$

The final free-precession period leads to refocusing of the S-spin chemical shifts and converts the antiphase operators to in-phase states which will be observable in the presence of I-spin decoupling:

$$
\begin{aligned}
&\mathrm{CH} \xrightarrow{\Omega_S\tau} \xrightarrow{\pi J_{I_1S}\tau\,(\tau=1/2J_{IS})} S_x \sin\beta, \\
&\mathrm{CH_2} \xrightarrow{\Omega_S\tau} \xrightarrow{\pi J_{I_1S}\tau\,(\tau=1/2J_{IS})} \\
&\qquad\quad \xrightarrow{\pi J_{I_2S}\tau\,(\tau=1/2J_{IS})} S_x \sin\beta\cos\beta, \\
&\mathrm{CH_3} \xrightarrow{\Omega_S\tau} \xrightarrow{\pi J_{I_1S}\tau\,(\tau=1/2J_{IS})} \\
&\qquad\quad \xrightarrow{\pi J_{I_2S}\tau\,(\tau=1/2J_{IS})} \\
&\qquad\quad \xrightarrow{\pi J_{I_3S}\tau\,(\tau=1/2J_{IS})} S_x \sin\beta\cos^2\beta.
\end{aligned}
\tag{4.20}
$$

Thus, the final ^{13}C spectrum will, in general, contain the resonances of all three types of carbon atoms and, as with INEPT, there will be a signal enhancement proportional to the ratio γ_I/γ_S. However, the different flip angle dependencies allow one to edit the spectrum according to multiplicity. If three spectra A, B, and C are recorded with flip angles of $\beta = 45°$, $90°$ and $135°$, respectively, then spectrum B contains only CH signals, the difference (A – C) has only CH_2 signals, while a spectrum containing only CH_3 signals corresponds to the linear combination $[(A + C)/\sqrt{2} - B]$.

5 Two-dimensional NMR

5.1 Introduction

A few of the more common two-dimensional NMR experiments will now be described using product operators. Our aim will be to see how they work and to explain their chief features with respect to phase and multiplet structure.

5.2 COSY

COSY: correlation spectroscopy. See OCP 32, Section 6.4 for illustrative COSY spectra and an attempt to explain the experiment's inner workings in terms of the vector model. The latter shows clearly why we need product operators to understand all but the most basic NMR experiments.

COSY (Fig. 5.1) is the archetypal two-dimensional NMR experiment. It was the first to be proposed (1971) and is still one of the most popular. It is used to correlate J-coupled spins for purposes of spectral assignment.

First consider a two-spin IS system. We will only calculate the evolution for spin I; S behaves identically, but with the labels I and S interchanged. Immediately after the second pulse we have:

$$\begin{aligned} I_z \xrightarrow{(90^\circ_x)_{\mathrm{IS}}} \xrightarrow{\Omega_{\mathrm{I}} t_1} \xrightarrow{\pi J_{\mathrm{IS}} t_1} & \\ & -I_y \cos\Omega_{\mathrm{I}} t_1 \cos\pi J_{\mathrm{IS}} t_1 + 2I_x S_z \cos\Omega_{\mathrm{I}} t_1 \sin\pi J_{\mathrm{IS}} t_1 \\ & + I_x \sin\Omega_{\mathrm{I}} t_1 \cos\pi J_{\mathrm{IS}} t_1 + 2I_y S_z \sin\Omega_{\mathrm{I}} t_1 \sin\pi J_{\mathrm{IS}} t_1 \\ \xrightarrow{(90^\circ_x)_{\mathrm{IS}}} & -I_z \cos\Omega_{\mathrm{I}} t_1 \cos\pi J_{\mathrm{IS}} t_1 - 2I_x S_y \cos\Omega_{\mathrm{I}} t_1 \sin\pi J_{\mathrm{IS}} t_1 \\ & + I_x \sin\Omega_{\mathrm{I}} t_1 \cos\pi J_{\mathrm{IS}} t_1 - 2I_z S_y \sin\Omega_{\mathrm{I}} t_1 \sin\pi J_{\mathrm{IS}} t_1 . \end{aligned} \tag{5.1}$$

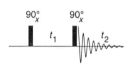

Fig. 5.1 The COSY pulse sequence.

Of the four terms, the first two (I_z and $2I_xS_y$) are not observable. The third (I_x) represents I magnetization that has evolved at frequency Ω_{I} in t_1 and will evolve at the same frequency in t_2. It therefore gives rise to a diagonal peak. The final term ($2I_zS_y$), on the other hand, represents S magnetization that has evolved at frequency Ω_{I} in t_1 but will evolve at frequency Ω_{S} in t_2. It therefore corresponds to a COSY cross peak between spins I and S. Its dependence on $\sin\pi J_{\mathrm{IS}} t_1$ shows that it will only be present if J_{IS} is non-zero.

We can see from the form of the operators I_x and $2I_zS_y$ that the diagonal peak will have an in-phase multiplet structure in F_2 while the cross peak will be antiphase with respect to the J_{IS} coupling. Also note that the two peaks will be 90° out-of-phase with respect to each other in F_2. Hence, if the cross peak is phased to be absorptive in this dimension, the diagonal peak will be dispersive. A schematic cross-section parallel to F_2 (Fig. 5.2) perhaps shows this more clearly.

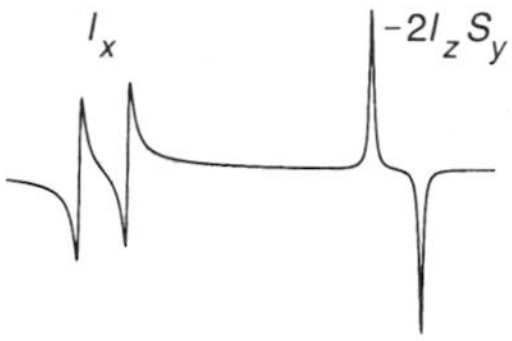

Fig. 5.2 Schematic cross-section, parallel to the F_2 axis, through a COSY spectrum of an IS spin system.

Inspection of the operator terms that exist at the start of acquisition in t_2 only gives us the form of the spectrum in the F_2 dimension. To determine the appearance of the COSY spectrum in the F_1 dimension we must examine the modulation that these observables have acquired during t_1. If we expand the amplitudes of I_x and $2I_zS_y$ in Eqn 5.1 using trigonometric relations we find:

$$
\begin{aligned}
&I_x \sin\Omega_I t_1 \cos\pi J_{IS}t_1 - 2I_zS_y \sin\Omega_I t_1 \sin\pi J_{IS}t_1 \\
&= I_x \tfrac{1}{2}\left[\sin(\Omega_I + \pi J_{IS})t_1 + \sin(\Omega_I - \pi J_{IS})t_1\right] \\
&- 2I_zS_y \tfrac{1}{2}\left[\cos(\Omega_I - \pi J_{IS})t_1 - \cos(\Omega_I + \pi J_{IS})t_1\right].
\end{aligned} \tag{5.2}
$$

$\sin(A \pm B) = \sin A \cos B \pm \cos A \sin B$
$\cos(A \pm B) = \cos A \cos B \mp \sin A \sin B$

The two frequencies $\Omega_I + \pi J_{IS}$ and $\Omega_I - \pi J_{IS}$ correspond to the two components of the I-spin doublet. In the I_x term they have the same sign and so represent an *in-phase* doublet while in the $2I_zS_y$ term they have opposite signs—an antiphase doublet. The fact that the respective modulations are sine and cosine indicates that in F_1 as well as F_2 there will be a 90° phase difference between the diagonal and cross peaks. Thus the multiplet structures and relative phases of the diagonal and cross peaks will be the same in both the F_1 and F_2 dimensions of a COSY spectrum. The appearance of the COSY spectrum of an IS spin system is therefore as shown schematically in Fig. 5.3. Finally, note that both diagonal and cross peaks are *amplitude* modulated which means that the procedure outlined in Section 2.4, or something equivalent, will be needed to determine the sense of precession during t_1.

A disadvantage of *antiphase* cross peaks is that the positive and negative portions tend to overlap and cancel one another. This can be troublesome for large molecules which have broad lines; in such cases TOCSY (Section 5.5) provides a useful alternative.

This analysis is easily extended to a three-spin ISR system. Starting with the I_z magnetization as before, and omitting all of the intermediate steps:

$$
\begin{aligned}
I_z &\xrightarrow{(90^\circ_x)_{ISR}} \xrightarrow{\Omega_I t_1} \xrightarrow{\pi J_{IS}t_1} \xrightarrow{\pi J_{IR}t_1} \xrightarrow{(90^\circ_x)_{ISR}} \\
&- I_z \cos\Omega_I t_1 \cos\pi J_{IS}t_1 \cos\pi J_{IR}t_1 \\
&- 2I_xS_y \cos\Omega_I t_1 \sin\pi J_{IS}t_1 \cos\pi J_{IR}t_1 \\
&+ I_x \sin\Omega_I t_1 \cos\pi J_{IS}t_1 \cos\pi J_{IR}t_1 \qquad (1) \\
&- 2I_zS_y \sin\Omega_I t_1 \sin\pi J_{IS}t_1 \cos\pi J_{IR}t_1 \qquad (2) \\
&- 2I_xR_y \cos\Omega_I t_1 \cos\pi J_{IS}t_1 \sin\pi J_{IR}t_1 \\
&+ 4I_zS_yR_y \cos\Omega_I t_1 \sin\pi J_{IS}t_1 \sin\pi J_{IR}t_1 \\
&- 2I_zR_y \sin\Omega_I t_1 \cos\pi J_{IS}t_1 \sin\pi J_{IR}t_1 \qquad (3) \\
&- 4I_xS_yR_y \sin\Omega_I t_1 \sin\pi J_{IS}t_1 \sin\pi J_{IR}t_1.
\end{aligned} \tag{5.3}
$$

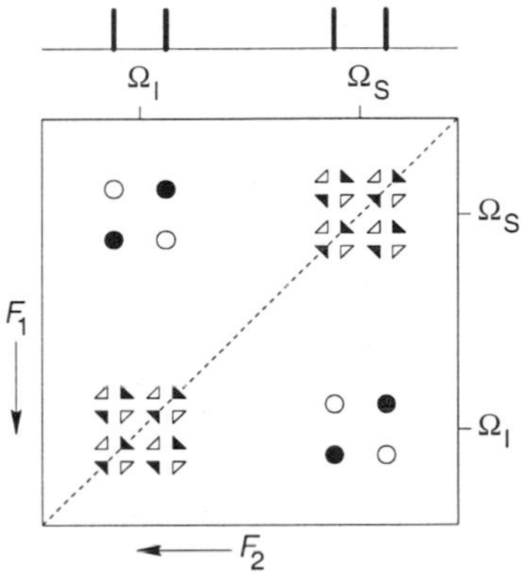

Fig. 5.3 Schematic COSY spectrum of an IS-spin system, showing dispersive, in-phase diagonal peaks and absorptive, antiphase cross peaks. Solid and open symbols represent positive and negative peaks, respectively. Circles represent the double absorption lineshape (Fig. 2.6a); the double dispersion lineshape (Fig. 2.6b) is indicated by four triangles.

Of the eight terms only those labelled (1), (2) and (3) give rise to observable signals. (1) Is a diagonal peak; in the F_2 dimension it is in-phase with respect to both J_{IS} and J_{IR} and dispersive (both of these because the operator is I_x), while in the F_1 dimension it is also in-phase with respect to both J_{IS} and J_{IR} (cosine modulated by both couplings) and dispersive (overall sine modulated). (2) Is an IS cross peak; in the F_2 dimension it is antiphase with respect to J_{IS} but in-phase with respect to J_{IR} and absorptive (both of these because the operator is $2I_zS_y$), while in the F_1 dimension it is also antiphase with respect to J_{IS} (sine modulated by J_{IS}) but in-phase with respect to J_{IR} (cosine modulated by J_{IR}) and absorptive (overall cosine modulated). Note that the antiphase structure arises from the *active* coupling (J_{IS} for the IS cross peak) while the *passive* coupling (J_{IR}) gives in-phase structure. (3) Is the analogous IR cross peak. Fig. 5.4 shows these three terms in the form of a (partial) schematic COSY spectrum.

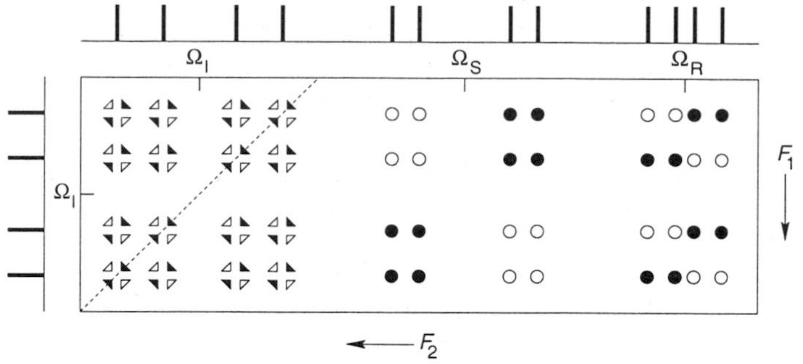

Fig. 5.4 Part of a schematic COSY spectrum of an ISR-spin system, showing the diagonal and cross peaks centred at frequency Ω_I in the F_1 dimension. The dashed line indicates the part of the diagonal that passes through this section of the spectrum.

5.3 DQF-COSY

DQF-COSY: double-quantum filtered COSY.

A drawback of COSY is that the diagonal peaks are rather intense because they are in phase and cover a large area of the two-dimensional NMR spectrum on account of their dispersive nature. Thus, cross peaks close to the diagonal can be obscured. A popular variant, DQF-COSY (Fig. 5.5), attempts to solve this problem by causing the diagonal peaks to be antiphase and *nearly* pure absorption. Any singlet resonances from isolated spins are also suppressed. As a result, DQF-COSY is often considered superior to the basic COSY experiment.

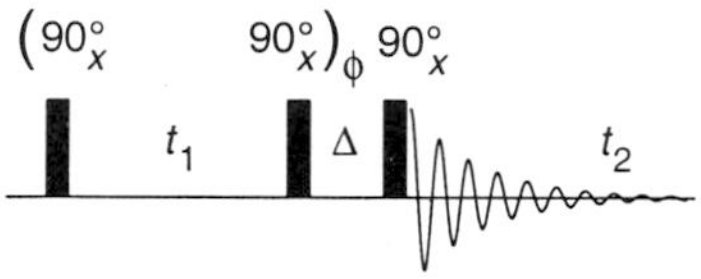

Fig. 5.5 The double-quantum filtered COSY pulse sequence.

The phase, ϕ, of the first two pulses is cycled as in the one-dimensional double-quantum filter (Section 4.3) to select only pure double-quantum coherence during the negligible interval, Δ. For the I-spin magnetization of an IS spin system we find:

$$I_z \xrightarrow{(90^\circ_x)_{IS}} \xrightarrow{\Omega_I t_1} \xrightarrow{\pi J_{IS} t_1} \xrightarrow{(90^\circ_x)_{IS}} -2I_xS_y \cos\Omega_I t_1 \sin\pi J_{IS} t_1 \tag{5.4}$$

where we have retained only the term containing the desired IS double-quantum coherence to be selected by the phase cycle. From Eqn 4.5, the product operator $2I_xS_y$ is equal to $DQ_y - ZQ_y$, the zero-quantum component of which will be removed by the phase cycle, so we can happily replace the $2I_xS_y$ by the operator $DQ_y = (2I_xS_y + 2I_yS_x)/2$, to obtain:

$$\begin{aligned} &\xrightarrow{\text{select DQ}} -\tfrac{1}{2}\left(2I_xS_y + 2I_yS_x\right)\cos\Omega_I t_1 \sin\pi J_{IS} t_1 \\ &\xrightarrow{(90^\circ_x)_{IS}} -\tfrac{1}{2}\left(2I_xS_z + 2I_zS_x\right)\cos\Omega_I t_1 \sin\pi J_{IS} t_1 . \end{aligned} \tag{5.5}$$

Both parts of this final state are observable. The first represents a diagonal peak (which precesses at Ω_I in both t_1 and t_2) that will be antiphase with respect to J_{IS} in F_2 (because the operator is $2I_xS_z$) *and* in F_1 (sine modulated by J_{IS}). The second represents the IS cross peak (precesses at Ω_I in F_1 and Ω_S in F_2) which will also be antiphase with respect to J_{IS} in both F_1 and F_2.

Finally, note that both diagonal and cross peak have the same (amplitude) modulation in t_1 and the same phase in t_2 and can therefore both be phased to pure absorption in both dimensions. The DQF-COSY spectrum of an IS spin system is therefore as shown in Fig. 5.6. The only disadvantage of DQF-COSY with respect to COSY is revealed by the factor of one-half in Eqn 5.5; the sensitivity of the double-quantum filtered experiment is half that of the basic experiment.

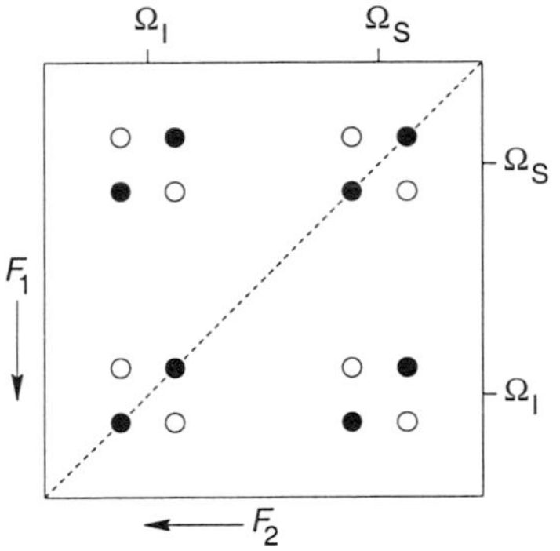

Fig. 5.6 Schematic DQF-COSY spectrum of an IS spin system, showing the absorptive, antiphase diagonal *and* cross peaks.

Another feature of DQF-COSY emerges when the product operator calculation is performed for an ISR spin system. After the second 90° pulse we find:

$$
\begin{aligned}
I_z &\xrightarrow{(90^\circ_x)_{\text{ISR}}} \xrightarrow{\Omega_I t_1} \xrightarrow{\pi J_{IS} t_1} \xrightarrow{\pi J_{IR} t_1} \xrightarrow{(90^\circ_x)_{\text{ISR}}} \\
&-2I_x S_y \cos\Omega_I t_1 \sin\pi J_{IS} t_1 \cos\pi J_{IR} t_1 \\
&-2I_x R_y \cos\Omega_I t_1 \cos\pi J_{IS} t_1 \sin\pi J_{IR} t_1 \\
&+4I_z S_y R_y \cos\Omega_I t_1 \sin\pi J_{IS} t_1 \sin\pi J_{IR} t_1
\end{aligned}
\tag{5.6}
$$

where again we have only retained the terms containing double-quantum coherence. As for the IS spin system, because of the use of a phase cycle, we can now replace these mixed operators with their pure double-quantum equivalents:

$$
\begin{aligned}
&\xrightarrow{\text{select DQ}} -\tfrac{1}{2}\left(2I_x S_y + 2I_y S_x\right)\cos\Omega_I t_1 \sin\pi J_{IS} t_1 \cos\pi J_{IR} t_1 \\
&\quad -\tfrac{1}{2}\left(2I_x R_y + 2I_y R_x\right)\cos\Omega_I t_1 \cos\pi J_{IS} t_1 \sin\pi J_{IR} t_1 \\
&\quad -\tfrac{1}{2}\left(4I_z S_x R_x - 4I_z S_y R_y\right)\cos\Omega_I t_1 \sin\pi J_{IS} t_1 \sin\pi J_{IR} t_1 \\
&\xrightarrow{(90^\circ_x)_{\text{ISR}}} -\tfrac{1}{2}\left(2I_x S_z + 2I_z S_x\right)\cos\Omega_I t_1 \sin\pi J_{IS} t_1 \cos\pi J_{IR} t_1 \\
&\quad -\tfrac{1}{2}\left(2I_x R_z + 2I_z R_x\right)\cos\Omega_I t_1 \cos\pi J_{IS} t_1 \sin\pi J_{IR} t_1 \\
&\quad +\tfrac{1}{2}\left(4I_y S_x R_x - 4I_y S_z R_z\right)\cos\Omega_I t_1 \sin\pi J_{IS} t_1 \sin\pi J_{IR} t_1 .
\end{aligned}
\tag{5.7}
$$

The first pair of product operators gives rise to an absorptive, antiphase (with respect to J_{IS}) I-spin diagonal peak and an absorptive, antiphase IS cross peak. The second pair will give rise to another absorptive, antiphase (but this time with respect to J_{IR}) contribution to the I-spin diagonal peak and an absorptive, antiphase IR cross peak. All this could have been extrapolated from the result for the IS spin system, but a novel feature is represented by the final pair of three-spin product operators. The first term of the pair is not observable but the second, $4I_yS_zR_z$, represents another contribution to the I-spin diagonal peak. It is doubly antiphase, however—antiphase with respect to both J_{IS} and J_{IR}—in both dimensions and is also *dispersive* in both dimensions. Thus, in systems of three or more spins, there *is* a dispersive contribution to the diagonal of a DQF-COSY. However, because it is antiphase with respect to two couplings, the cancellation of positive and negative intensities means that it is normally weaker than the absorptive components, and the diagonal peak multiplets remain largely absorptive.

5.4 NOESY

NOESY: nuclear Overhauser effect (or enhancement) spectroscopy.

Nuclear Overhauser effect, NOE, is the name given to the transfer of z magnetization from one spin to another by cross relaxation induced by the dipolar interaction of the two spins. It gives information on the 'through-space' separation of the nuclei, see OCP 32, Section 5.4.

NOESY (Fig. 5.7) is one of the oldest two-dimensional NMR experiments and still one of the most widely used. In contrast to COSY, which correlates J-coupled spins, the existence of a NOESY cross peak indicates a *nuclear Overhauser effect* between the two nuclei, implying that they can be no more than a few-hundred picometres apart.

Consider a two-spin IS system. Starting with z magnetization of spin I, the effect of the first part of the pulse sequence is identical to COSY (Eqn 5.1):

$$\begin{aligned} I_z &\xrightarrow{(90^\circ_x)_{\mathrm{IS}}} \xrightarrow{\Omega_\mathrm{I} t_1} \xrightarrow{\pi J_{\mathrm{IS}} t_1} \xrightarrow{(90^\circ_x)_{\mathrm{IS}}} \\ & - I_z \cos \Omega_\mathrm{I} t_1 \cos \pi J_{\mathrm{IS}} t_1 - 2 I_x S_y \cos \Omega_\mathrm{I} t_1 \sin \pi J_{\mathrm{IS}} t_1 \\ & + I_x \sin \Omega_\mathrm{I} t_1 \cos \pi J_{\mathrm{IS}} t_1 - 2 I_z S_y \sin \Omega_\mathrm{I} t_1 \sin \pi J_{\mathrm{IS}} t_1 . \end{aligned} \tag{5.8}$$

The full meaning of the terms coherence and coherence order are explained in Chapters 6 and 9.

The phase cycling is designed to select only coherence order $p = 0$ during τ_m, i.e. populations and zero-quantum coherences.

$$\xrightarrow{\text{select } p=0} -I_z \cos \Omega_\mathrm{I} t_1 \cos \pi J_{\mathrm{IS}} t_1 + ZQ_y \cos \Omega_\mathrm{I} t_1 \sin \pi J_{\mathrm{IS}} t_1 . \tag{5.9}$$

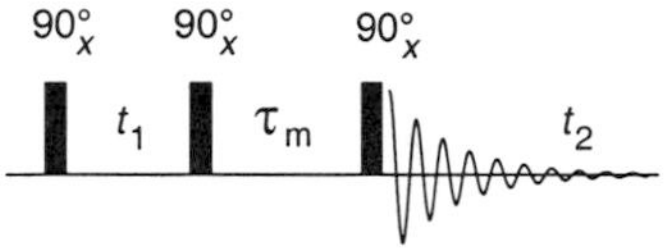

Fig. 5.7 The NOESY pulse sequence. In this, and all subsequent pulse sequences, the use of phase cycling is *not* indicated (cf. Fig. 5.5).

During the mixing period, τ_m, z magnetization is passed from I to S either by cross relaxation (i.e. NOE) or chemical exchange. In general terms, the mixing period results in the transformation

$$I_z \longrightarrow a I_z + b S_z \tag{5.10}$$

where the coefficients a and b are numbers that depend on τ_m and the efficiency of the polarization transfer. Appendix 5.1 gives some insight into the origin of this effect. Remembering that the zero-quantum coherence in Eqn 5.9 precesses during τ_m, we find:

The eagle-eyed will have spotted that NOESY (Fig. 5.7) and DQF-COSY (Fig. 5.5) have seemingly identical pulse sequences. They give different spectra because of their different phase cycles which select z magnetization and double-quantum coherence respectively during the second delay (τ_m or Δ). See Chapter 6 for more details.

$$\begin{aligned} \xrightarrow{(\Omega_\mathrm{I}+\Omega_\mathrm{S})\tau_\mathrm{m}} & -a I_z \cos \Omega_\mathrm{I} t_1 \cos \pi J_{\mathrm{IS}} t_1 - b S_z \cos \Omega_\mathrm{I} t_1 \cos \pi J_{\mathrm{IS}} t_1 \\ & + ZQ_y \cos \Omega_\mathrm{I} t_1 \sin \pi J_{\mathrm{IS}} t_1 \cos(\Omega_\mathrm{I} - \Omega_\mathrm{S})\tau_\mathrm{m} \\ & - ZQ_x \cos \Omega_\mathrm{I} t_1 \sin \pi J_{\mathrm{IS}} t_1 \sin(\Omega_\mathrm{I} - \Omega_\mathrm{S})\tau_\mathrm{m} , \end{aligned} \tag{5.11}$$

and after the final pulse:

$$\begin{aligned} \xrightarrow{(90^\circ_x)_{\mathrm{IS}}} & a I_y \cos \Omega_\mathrm{I} t_1 \cos \pi J_{\mathrm{IS}} t_1 + b S_y \cos \Omega_\mathrm{I} t_1 \cos \pi J_{\mathrm{IS}} t_1 \\ & + \tfrac{1}{2}\left(2 I_z S_x - 2 I_x S_z\right) \cos \Omega_\mathrm{I} t_1 \sin \pi J_{\mathrm{IS}} t_1 \cos(\Omega_\mathrm{I} - \Omega_\mathrm{S})\tau_\mathrm{m} \\ & - \tfrac{1}{2}\left(2 I_x S_x - 2 I_z S_z\right) \cos \Omega_\mathrm{I} t_1 \sin \pi J_{\mathrm{IS}} t_1 \sin(\Omega_\mathrm{I} - \Omega_\mathrm{S})\tau_\mathrm{m} . \end{aligned} \tag{5.12}$$

NOE cross peaks *only* appear if there is an exchange of z magnetization during τ_m, i.e. $b \neq 0$ in Eqn 5.10.

The first term, $a I_y$, is an in-phase diagonal peak which is absorptive in both dimensions. The second, $b S_y$, is an in-phase IS cross peak, also absorptive in both dimensions. The third, $2 I_z S_x - 2 I_x S_z$, represents the unwanted zero-quantum diagonal and cross peaks which are antiphase and dispersive in both F_1 and F_2 dimensions. These peaks can be suppressed by recording and adding together several NOESY spectra with a range of values of the mixing time τ_m. Alternatively, in the spectra of large molecules, they can often be ignored because the zero-quantum relaxation rate is very fast and the antiphase contributions tend to cancel. In any case, they do not affect the integrated volume of the cross peaks. The fourth term is not observable.

5.5 TOCSY

TOCSY (Fig. 5.8) is a popular alternative to COSY and DQF-COSY for large biological molecules. This is because it yields in-phase cross peaks which, unlike the antiphase cross peaks in COSY, do not tend to disappear as a result of the large linewidths encountered for molecules of high molecular mass. Another property of TOCSY is that it correlates remotely connected spins as well as directly connected spins and can thus be used to map out entire spin systems. The key feature of the TOCSY experiment is that it uses a period of *spin locking* to achieve coherence transfer.

TOCSY: total correlation spectroscopy.

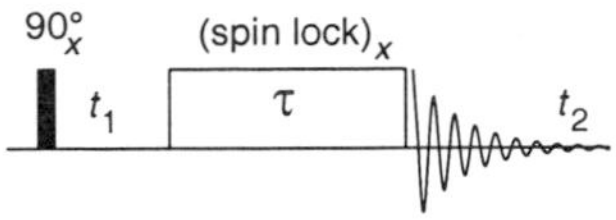

Fig. 5.8 The TOCSY pulse sequence.

In its simplest form, the spin locking field is just a long, strong radiofrequency pulse along a specified axis in the rotating frame. If $\omega_1 = -\gamma B_1$ is much larger than all the offset frequencies, Ω, precession around the z axis is suppressed, all chemical shift differences become irrelevant and the spins effectively become *equivalent*.

It needs a small amount of quantum mechanics to work out exactly how the various product operators evolve under the influence of the spin locking field (Chapter 9). The results for an IS spin system, with the spin locking field along the x axis, are that the sum of the x magnetizations of the two spins does not evolve, while their difference does:

$$\begin{aligned} I_x + S_x &\xrightarrow{\textit{spin lock}} I_x + S_x \\ I_x - S_x &\xrightarrow{\textit{spin lock}} (I_x - S_x)\cos 2\pi J_{IS}\tau + (2I_yS_z - 2I_zS_y)\sin 2\pi J_{IS}\tau. \end{aligned} \tag{5.13}$$

These equations ignore the effects of relaxation during the relatively long spin lock period. In particular, cross relaxation will give rise to *rotating frame Overhauser effects* (ROEs). The TOCSY experiment may be performed in a manner designed to maximize ROEs while minimizing the effects of evolution under the J coupling; in this case it is referred to as ROESY.

The y and z components evolve in an analogous manner, but also precess around the x axis at frequency ω_1. Note that the rate of evolution of in-phase into antiphase terms under the influence of spin locking is $2\pi J_{IS}$ rather than the πJ_{IS} arising from free precession.

The evolution of the magnetization of a single spin, say I_x, can therefore be calculated as follows:

$$\begin{aligned} I_x &= \tfrac{1}{2}(I_x + S_x) + \tfrac{1}{2}(I_x - S_x) \\ &\xrightarrow{\textit{spin lock}} I_x\tfrac{1}{2}(1+\cos 2\pi J_{IS}\tau) + S_x\tfrac{1}{2}(1-\cos 2\pi J_{IS}\tau) \\ &\qquad + (2I_yS_z - 2I_zS_y)\tfrac{1}{2}\sin 2\pi J_{IS}\tau. \end{aligned} \tag{5.14}$$

Following only the I magnetization, the state of the IS spin system at the end of the t_1 period can be written:

$$\begin{aligned} I_z &\xrightarrow{(90^\circ_x)_{IS}} \xrightarrow{\Omega_I t_1} \xrightarrow{\pi J_{IS} t_1} \\ &-I_y \cos\Omega_I t_1 \cos\pi J_{IS}t_1 + 2I_xS_z \cos\Omega_I t_1 \sin\pi J_{IS}t_1 \\ &+ I_x \sin\Omega_I t_1 \cos\pi J_{IS}t_1 + 2I_yS_z \sin\Omega_I t_1 \sin\pi J_{IS}t_1. \end{aligned} \tag{5.15}$$

This process is analogous to the dephasing of transverse magnetization in an inhomogeneous B_0 field.

Now the first two of these terms will dephase during the spin locking if the B_1 field is spatially inhomogeneous (which it certainly will be) and along the same axis (x) as the initial 90° pulse. They can therefore be neglected. The fourth term, perhaps surprisingly, does not dephase and will give rise to an antiphase, dispersive contribution to the two-dimensional lineshapes. It is the third term that represents the desired component of the magnetization, as it will transfer into S_x during the spin locking:

N.B. $1 + \cos 2A = 2\cos^2 A$ and $1 - \cos 2A = 2\sin^2 A$

$$\begin{aligned} I_x \sin\Omega_I t_1 \cos\pi J_{IS} t_1 &\xrightarrow{\text{spin lock}} \\ & I_x \sin\Omega_I t_1 \cos\pi J_{IS} t_1 \cos^2\pi J_{IS}\tau \\ & + S_x \sin\Omega_I t_1 \cos\pi J_{IS} t_1 \sin^2\pi J_{IS}\tau \\ & + \left(2I_y S_z - 2I_z S_y\right)\sin\Omega_I t_1 \cos\pi J_{IS} t_1 \tfrac{1}{2}\sin 2\pi J_{IS}\tau. \end{aligned} \tag{5.16}$$

The antiphase term ($2I_yS_z - 2I_zS_y$) can be suppressed by adding together several experiments with different values of the mixing time, τ. This works because $\sin 2\pi J_{IS}\tau$ changes sign as a function of τ while $\cos^2\pi J_{IS}\tau$ and $\sin^2\pi J_{IS}\tau$ do not. Both diagonal (I_x) and cross peaks (S_x) are in-phase in both dimensions and, since both have the same (amplitude) modulation, can both be phased to pure absorption (Fig. 5.9).

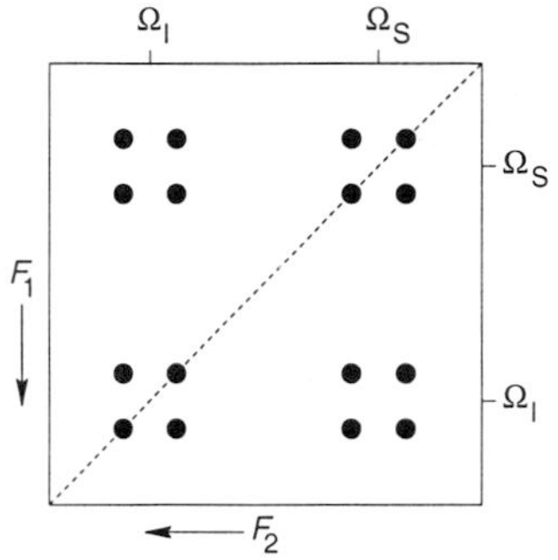

Fig. 5.9 Schematic TOCSY spectrum of an IS spin system, showing all peaks in positive absorption.

5.6 HMQC

HMQC: heteronuclear multiple-quantum correlation

All the two-dimensional NMR experiments considered so far have been homonuclear correlation techniques. As an example of a *heteronuclear* correlation technique we will now perform a product operator analysis of the basic HMQC experiment (Fig. 5.10) for correlating protons and heteronuclei such as ^{13}C or ^{15}N through their J couplings.

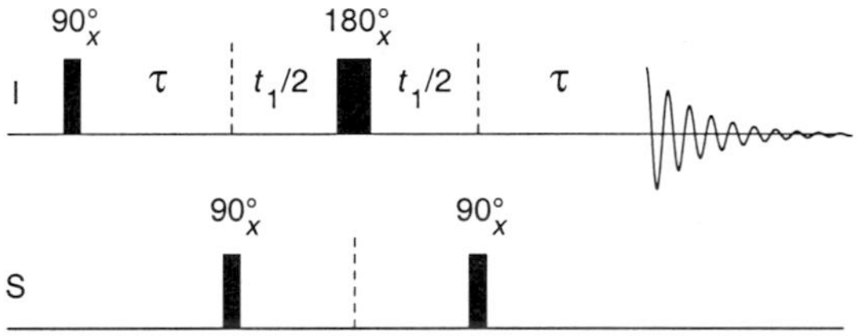

Fig. 5.10 The HMQC pulse sequence.

The essence of a so-called inverse experiment is that sensitivity is maximized by observing the free induction decay of the more sensitive nucleus (typically ^1H). Starting with z magnetization of spin I, we find:

$$\begin{aligned} I_z &\xrightarrow{(90^\circ_x)_I} \xrightarrow{\Omega_I\tau} \xrightarrow{\pi J_{IS}\tau} \\ & -I_y \cos\Omega_I\tau\cos\pi J_{IS}\tau + 2I_xS_z\cos\Omega_I\tau\sin\pi J_{IS}\tau \\ & + I_x \sin\Omega_I\tau\cos\pi J_{IS}\tau + 2I_yS_z\sin\Omega_I\tau\sin\pi J_{IS}\tau. \end{aligned} \tag{5.17}$$

The fixed free precession interval τ is set to $1/2J_{IS}$ so that the antiphase terms are maximized:

$$
\begin{aligned}
&= 2I_xS_z\cos\Omega_I\tau + 2I_yS_z\sin\Omega_I\tau \\
&\xrightarrow{(90^\circ_x)_S} -2I_xS_y\cos\Omega_I\tau - 2I_yS_y\sin\Omega_I\tau.
\end{aligned}
\tag{5.18}
$$

These two terms both represent mixtures of heteronuclear zero- and double-quantum coherence. In this experiment, phase cycling is *not* used to separate the two. Note that the phase of these coherences depends on $\Omega_I\tau$ and, at this point in the experiment, it is not clear how pure absorption lineshapes are going to be obtained. These heteronuclear multiple-quantum coherences now evolve during the t_1 period of the experiment. As a short-cut in the calculation we can note that the 180° pulse on the I spin refocuses the I-spin chemical shifts during the evolution period. Therefore we need only calculate the evolution under the S-spin chemical shift together with the inversion of I_y by the 180° pulse:

Remember (Section 4.3) that zero- and double-quantum coherence do not evolve under the IS J coupling.

$$
\begin{aligned}
&\xrightarrow{(180^\circ_x)_I} -2I_xS_y\cos\Omega_I\tau + 2I_yS_y\sin\Omega_I\tau \\
&\xrightarrow{\Omega_S t_1} -2I_xS_y\cos\Omega_I\tau\cos\Omega_S t_1 + 2I_xS_x\cos\Omega_I\tau\sin\Omega_S t_1 \\
&\qquad + 2I_yS_y\sin\Omega_I\tau\cos\Omega_S t_1 - 2I_yS_x\sin\Omega_I\tau\sin\Omega_S t_1.
\end{aligned}
\tag{5.19}
$$

Thus, the only evolution during t_1 occurs at the chemical shift frequency of spin S. The heteronuclear multiple-quantum coherences are now reconverted into I-spin single-quantum coherence:

$$
\xrightarrow{(90^\circ_x)_S} -2I_xS_z\cos\Omega_I\tau\cos\Omega_S t_1 + 2I_yS_z\sin\Omega_I\tau\cos\Omega_S t_1 \tag{5.20}
$$

where we have retained only the terms that lead to observable signals. These antiphase coherences evolve under the influence of the spin I chemical shift and the coupling J_{IS} during the final $\tau = 1/2J_{IS}$ interval prior to acquisition:

$$
\begin{aligned}
&\xrightarrow{\Omega_I\tau} -2I_xS_z\cos^2\Omega_I\tau\cos\Omega_S t_1 - 2I_yS_z\cos\Omega_I\tau\sin\Omega_I\tau\cos\Omega_S t_1 \\
&\qquad + 2I_yS_z\cos\Omega_I\tau\sin\Omega_I\tau\cos\Omega_S t_1 - 2I_xS_z\sin^2\Omega_I\tau\cos\Omega_S t_1 \\
&\quad = -2I_xS_z\cos\Omega_S t_1 \\
&\xrightarrow{\pi J_{IS}\tau} -I_y\cos\Omega_S t_1.
\end{aligned}
\tag{5.21}
$$

The chemical shift evolution during the first τ period has been refocused in the second. Thus, pure absorption lineshapes can be obtained. The final multiplet structure of the peaks in the two-dimensional spectrum is a singlet in F_1 and an in-phase doublet in F_2. Broadband decoupling of the S spin can be used to obtain singlets in F_2.

5.7 HSQC

As a method of J-correlating heteronuclei, HMQC has largely been superseded by HSQC. This technique, which turns out to be much simpler to understand, is the basis of many more complex *three-dimensional* NMR

HSQC: heteronuclear single-quantum correlation.

experiments which play a central role in modern NMR studies of biomolecules such as proteins.

A simple HSQC pulse sequence is shown in Fig. 5.11. To break the monotony of analysing pulse sequences step-by-step, we treat this experiment as being built up from a small number of basic, and by now familiar elements. Knowing the behaviour of each of the building blocks, we can combine them in order to understand the whole sequence. This approach is almost essential when designing more complex experiments, which are developed by combining standard elements in interesting ways.

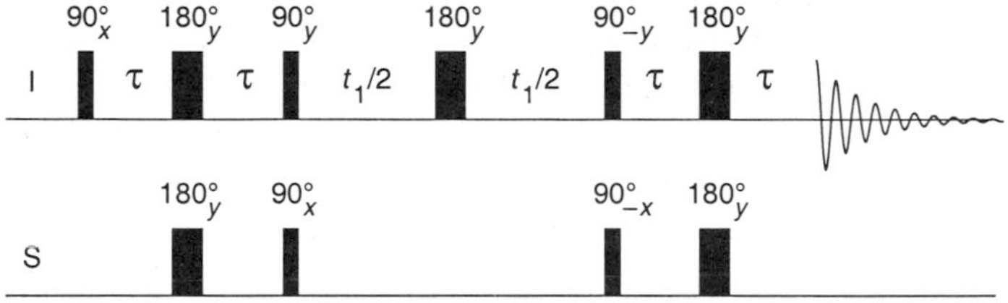

Fig. 5.11 The HSQC pulse sequence.

The first element of the HSQC experiment should be familiar as an INEPT sequence, which we have already analysed in considerable detail (see Section 4.2). At the end of this block the state of a two-spin IS system is (Eqn 4.3)

$$a2I_zS_y + bS_y, \tag{5.22}$$

where a and b are the inherent amplitudes of the I- and S-spin magnetizations, respectively. This is followed by a t_1 evolution period, in the middle of which a 180° pulse is applied to spin I. The evolution arising from the J coupling of I and S will be refocused, leaving only evolution under the S-spin chemical shift together with the effects of the 180° pulse on spin I. Thus, at the end of the evolution period the state of the system will be

$$-a2I_zS_y\cos\Omega_St_1 + bS_y\cos\Omega_St_1 + a2I_zS_x\sin\Omega_St_1 - bS_x\sin\Omega_St_1. \tag{5.23}$$

The final element of HSQC is simply a reversal of the initial INEPT sequence (that is, all the pulses and delays applied in reverse order, with the phase of the 90° pulses inverted), except that the first pulse has been removed. Using Eqns 4.1 and 4.3 in reverse, we can immediately deduce that the sequence will effect the two transformations

$$2I_zS_y \to -I_y, \qquad S_y \to S_z. \tag{5.24}$$

Thus, the evolution of the first two terms in Eqn 5.23 can be easily calculated. The pulse sequence is followed by observation of the NMR signal of spin I, so that only the first term will give rise to any observable signal. The remaining two terms in Eqn 5.23 are slightly more complicated, but neither results in an observable I-spin signal.

The third term in Eqn 5.23 is converted into (unobservable) zero- and double-quantum coherences, while the final term becomes antiphase magnetization on spin S.

Combining these results, the observable I-spin magnetization at the end of the HSQC pulse sequence is

$$aI_y\cos\Omega_St_1. \tag{5.25}$$

As with HMQC (Eqn 5.21) the signal from the I spin has been labelled with a modulation arising from the S spin, and then returned to I for detection. Modifying the relative phases of the INEPT and reverse INEPT sequences allows detection of a sine-modulated component, and so the sign of Ω_S can be determined.

5.8 Three- and four-dimensional NMR

NMR is routinely used to study large proteins and other biomolecules, and for many of these systems two-dimensional NMR techniques do not provide enough information. It is common to use three-dimensional, and even four-dimensional NMR experiments, which allow three (or four) different frequencies to be correlated (see Clore and Gronenborn 1994). There is a huge variety of such experiments, and it would take most of this book to analyse even the most common ones in detail. Fortunately, however, it is not necessary to perform such an analysis, as many of them can be easily understood by combining the basic building blocks we have already seen.

One important group of experiments combines an HSQC sequence with a homonuclear technique such as NOESY or TOCSY. This can be achieved by replacing the first 90° pulse of the homonuclear experiment with an HSQC sequence. Similarly, it is possible to correlate the frequencies of three different heteronuclei (e.g. ^{1}H, ^{15}N and ^{13}C nuclei) by nesting one HSQC sequence inside another.

Appendix 5.1 NOE, cross relaxation, and the Solomon equations

The important part of a NOESY experiment—the bit when things really happen—is the mixing period, τ_m. During this interval, cross relaxation and/or slow chemical exchange transfers some of the intensity of the modulated I_z state into S_z, carrying with it the modulation at frequencies Ω_I and πJ_{IS}. We can see how cross relaxation brings about this transfer by considering the energy level diagram for a two-spin IS system (Fig. 5.12).

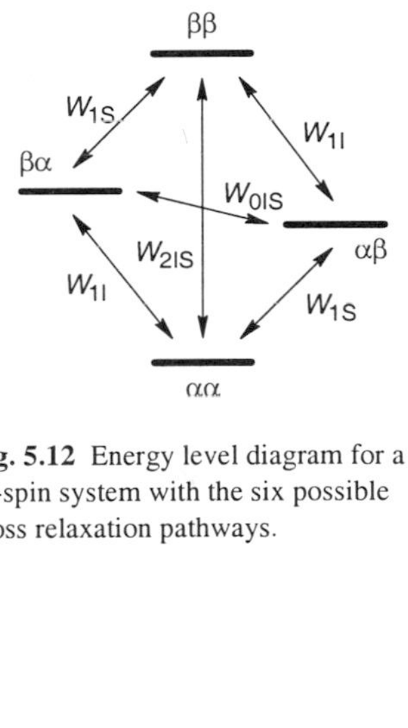

Fig. 5.12 Energy level diagram for a IS-spin system with the six possible cross relaxation pathways.

Spin–lattice or longitudinal relaxation is caused by fluctuating magnetic fields in a liquid sample which induce transitions between spin states. If these fields arise from the dipolar interaction of the two spins, then all six possible transitions occur. The relaxation rate constants are labelled W_{1I} if the I spin flips and the magnetic quantum number m changes by $\Delta m = \pm 1$, W_{1S} if the S spin flips and $\Delta m = \pm 1$, W_{2IS} if both I and S flip in the same sense such that $\Delta m = \pm 2$, and W_{0IS} if both I and S flip in opposite senses and $\Delta m = 0$.

The effect of these spin flips is to move the populations of the four energy levels towards equilibrium. Ignoring the equilibrium population differences for simplicity for the moment, we have the rate equations:

$$\frac{dn_{\alpha\alpha}}{dt} = -\left(W_{1I} + W_{1S} + W_{2IS}\right)n_{\alpha\alpha} + W_{1I}n_{\beta\alpha} + W_{1S}n_{\alpha\beta} + W_{2IS}n_{\beta\beta}$$
$$\frac{dn_{\alpha\beta}}{dt} = -\left(W_{1I} + W_{1S} + W_{0IS}\right)n_{\alpha\beta} + W_{1I}n_{\beta\beta} + W_{1S}n_{\alpha\alpha} + W_{0IS}n_{\beta\alpha} \qquad (5.26)$$

$n_{\alpha\alpha}$, $n_{\alpha\beta}$, $n_{\beta\alpha}$, and $n_{\beta\beta}$ are the populations of the four energy levels shown in Fig. 5.12.

and similarly for $dn_{\beta\alpha}/dt$ and $dn_{\beta\beta}/dt$. After some tedious but straightforward rearrangement, one can obtain the two coupled rate equations

$$\begin{aligned}\frac{d(n_{\alpha\alpha}+n_{\alpha\beta}-n_{\beta\alpha}-n_{\beta\beta})}{dt} &= -\rho_I\left(n_{\alpha\alpha}+n_{\alpha\beta}-n_{\beta\alpha}-n_{\beta\beta}\right) \\ &\quad -\sigma_{IS}\left(n_{\alpha\alpha}-n_{\alpha\beta}+n_{\beta\alpha}-n_{\beta\beta}\right) \\ \frac{d(n_{\alpha\alpha}-n_{\alpha\beta}+n_{\beta\alpha}-n_{\beta\beta})}{dt} &= -\sigma_{IS}\left(n_{\alpha\alpha}+n_{\alpha\beta}-n_{\beta\alpha}-n_{\beta\beta}\right) \\ &\quad -\rho_S\left(n_{\alpha\alpha}-n_{\alpha\beta}+n_{\beta\alpha}-n_{\beta\beta}\right)\end{aligned} \tag{5.27}$$

where

$$\begin{aligned}\rho_I &= W_{0IS}+2W_{1I}+W_{2IS} \\ \rho_S &= W_{0IS}+2W_{1S}+W_{2IS} \\ \sigma_{IS} &= W_{2IS}-W_{0IS}.\end{aligned} \tag{5.28}$$

But $(n_{\alpha\alpha} + n_{\alpha\beta} - n_{\beta\alpha} - n_{\beta\beta})$ is just the total number of I spins in the α state minus the total number in the β state; $(n_{\alpha\alpha} + n_{\alpha\beta} - n_{\beta\alpha} - n_{\beta\beta})$ is therefore proportional to the z component of the I-spin magnetization. Similarly, $(n_{\alpha\alpha} - n_{\alpha\beta} + n_{\beta\alpha} - n_{\beta\beta})$ is proportional to the z component of the S-spin magnetization. Defining $a_I(t)$ and $a_S(t)$ as the amplitudes of the I_z and S_z operators at time t, we get

$$\begin{aligned}\frac{da_I}{dt} &= -\rho_I a_I - \sigma_{IS} a_S \\ \frac{da_S}{dt} &= -\sigma_{IS} a_I - \rho_S a_S.\end{aligned} \tag{5.29}$$

To ensure that relaxation restores the (Boltzmann) equilibrium magnetizations, we can simply replace a_I and a_S on the right-hand side by $(a_I - a_I^0)$ and $(a_S - a_S^0)$, where the superscript zero indicates the value at equilibrium:

This would not have been necessary if we had included the equilibrium populations at the outset.

$$\begin{aligned}\frac{da_I}{dt} &= -\rho_I\left(a_I - a_I^0\right) - \sigma_{IS}\left(a_S - a_S^0\right) \\ \frac{da_S}{dt} &= -\sigma_{IS}\left(a_I - a_I^0\right) - \rho_S\left(a_S - a_S^0\right).\end{aligned} \tag{5.30}$$

Solution of these coupled equations gives the time dependence of the coefficients $a_I(t)$ and $a_S(t)$, and hence the evolution of the z magnetization during the NOESY mixing period. The expressions are known as the Solomon equations. They show that z magnetization will be transferred between the two spins *provided* that the difference between the two cross relaxation rates, σ_{IS}, is not zero.

The Solomon equations may also be used to understand the nuclear Overhauser effect more generally. For example, the form of the σ_{IS} term (Eqn 5.28), without which there would be no magnetization transfer, shows clearly why the sign of the NOE depends on whether W_{2IS} is faster or slower than W_{0IS} (see OCP 32).

6 Phase cycling and pulsed field gradients

6.1 Introduction

Radiofrequency pulses are usually less discriminating than one might wish. Desired coherences are generally accompanied by unwanted ones which tend to clutter up the spectrum and obscure the signals of interest. One method of suppressing these intruders is *phase cycling*, a few examples of which we have already encountered. A two-step cycle was used in the INEPT experiment (Section 4.2) to select only the enhanced S-spin magnetization; four steps were used to ensure the correct operation of the double-quantum filter (Section 4.3). An alternative, and increasingly popular method is to use *pulsed field gradients* (i.e. spatially inhomogeneous magnetic fields) to dephase and later rephase only the required coherences. Phase cycling and/or pulsed field gradients are essential parts of any modern NMR experiment; the purpose of this chapter is to describe how they work.

6.2 Coherence transfer pathways

Table 6.1. Examples of some product operators and their coherence orders, p.

Operators	p
S_z, $2I_zS_z$, $4I_zS_zR_z$, $(I_xS_x + I_yS_y)$, $(2I_zS_yR_x - 2I_zS_xR_y)$	0
I_x, I_y, $2I_zS_y$, $4I_xS_zR_z$	±1
$(I_xS_x - I_yS_y)$, $(2I_zS_xR_y + 2I_zS_yR_x)$	±2

Every product operator can be assigned one or more *coherence orders*. Pure single-, double- and triple-quantum operators have coherence orders $p = \pm 1$, ± 2, ± 3, respectively, while population and zero-quantum operators have $p = 0$. Some examples are given in Table 6.1. Thus, a two-spin product operator such as $2I_xS_y$ has coherence orders $p = \pm 2$ and $p = 0$, as it represents a superposition of double- and zero-quantum coherences.

The significance of the ± signs can be seen by considering the example of the single quantum operator I_x ($p = \pm 1$). This can be written as

$$I_x = \tfrac{1}{2}(I_x + iI_y) + \tfrac{1}{2}(I_x - iI_y). \tag{6.1}$$

where the two (counter-rotating) terms on the right-hand side have coherence orders $p = -1$ and $+1$ respectively. Other $p = \pm n$ operators can be similarly decomposed into their $+n$ and $-n$ constituents.

During the free precession intervals of an NMR experiment the spectroscopist must have control over which coherence orders are present. For example, during the t_1 period of a DQF-COSY experiment one wishes to have only single-quantum ($p = \pm 1$) coherences evolving, but during the very brief delay that follows, only double-quantum ($p = \pm 2$) coherences are wanted. For every NMR experiment we can draw a *coherence level diagram* or *coherence transfer pathway diagram* which shows the *desired* coherence orders present in each free precession interval of the pulse sequence. The coherence transfer pathway diagram for a DQF-COSY experiment, for example, is shown in Fig. 6.1.

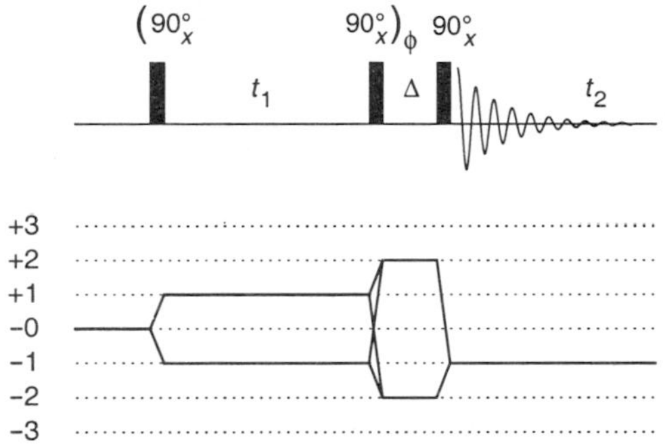

Fig. 6.1 Pulse sequence and coherence transfer pathway diagram for the double-quantum filtered COSY experiment.

The following features—general to all NMR experiments—should be noted:

The coherence transfer pathway starts at $p = 0$. The relatively long relaxation delay that precedes the pulse sequence ensures that the spin system is initially at thermal equilibrium, with only z magnetization present.

The coherence transfer pathway finishes at $p = -1$. Observable magnetization is always single-quantum coherence ($p = \pm 1$) and, if quadrature detection is used (which it ought to be), the x and y components are detected simultaneously and used as the real and imaginary inputs to a complex Fourier transform (Section 2.2). Purely by definition, this combination is taken to be I_x *plus* iI_y, which we have just identified as coherence order $p = -1$.

During the free precession intervals (t_1 and Δ in this case) prior to the acquisition period *both $p = +n$ and $p = -n$ coherences are shown as present.* When free precession is interrupted by a 90° pulse we get an *amplitude-modulated* signal (see Chapter 2) which can be thought of as the sum of two counter-rotating components; it is these that are represented by (for example) the pathways with $p = +1$ and -1 during the evolution period t_1.

The coherence order does not change during free precession. Only pulses can change the coherence order. A 90° pulse will in general change a coherence order p into all possible coherence orders allowed within the spin system; a perfect 180° pulse simply changes a coherence order p into $-p$.

The coherence transfer pathway diagram only shows the desired coherence transfer pathways during the pulse sequence. In DQF-COSY for example, if the first pulse is not precisely 90° then some z magnetization ($p = 0$) will be left during the t_1 period. Similarly, the second 90° pulse will excite $p = 0, \pm 1$, (and if the spin system is big enough), ± 3, ± 4, etc. All such extraneous coherence orders, which must be suppressed by phase cycling and/or pulsed field gradients, are excluded from the diagram.

The beauty of coherence transfer pathway diagrams is that they give immediate insight into what is actually going on during a pulse sequence. As we shall see, they allow us to arrive quickly at a simple recipe for cycling phases or inserting pulsed field gradients.

6.3 Phase cycling

What one might term the *First Rule of Phase Cycling* can be expressed as follows:

If the phase of a pulse or a group of pulses is shifted by ϕ, then a coherence undergoing a change in coherence order Δp experiences a phase shift $-\phi\Delta p$.

The desired change in coherence order Δp for any given pulse can be found simply by inspection of the coherence transfer pathway diagram. The corresponding pathway can then be selected by making the receiver phase ϕ^{R} follow the phase of the coherence: $\phi^{R} = -\phi\Delta p$. Other coherence transfer pathways will experience different phase shifts and should sum to zero over the phase cycle. For example, in DQF-COSY (Fig. 6.1) the final 90° pulse must change the coherence order by $\Delta p = -3$ (i.e. $+2 \rightarrow -1$) *and* by $\Delta p = +1$ (i.e. $-2 \rightarrow -1$). If the phase of the pulse is cycled through the values

$$\phi = 0°, 90°, 180°, 270° \qquad (6.2)$$

then, taking the pathway $\Delta p = -3$ first, we find that the receiver phase should go as

$$\phi^{R} = +3\phi = 0°, 270°, 540°, 810° \qquad (6.3)$$

which is equivalent to

$$\phi^{R} = 0°, 270°, 180°, 90°. \qquad (6.4)$$

On the other hand, if the pathway $\Delta p = +1$ is considered then the receiver phase cycle is

$$\phi^{R} = -\phi = 0°, -90°, -180°, -270° \qquad (6.5)$$

which is identical to Eqn 6.4. This simple four-step phase cycle thus selects both the pathways $\Delta p = -3$ and $+1$ as desired.

But how did we know that we had to use a *four*-step phase cycle (i.e. increment the pulse phase ϕ by 90° each time)? Why not three-, five- or six-steps? Although the above rule tells us whether particular pathways survive, how can we be sure that the undesirable pathways are being suppressed? The answers to these questions are provided by the *Second Rule of Phase Cycling*:

If a phase cycle uses steps of 360°/N then, along with the desired pathway Δp, pathways $\Delta p \pm nN$, where n = 1, 2, 3, ..., will also be selected. All other pathways are suppressed.

One way to use this rule is to write out all possible changes in the coherence order that can occur as a result of the pulse (Table 6.2, column A). In our DQF-COSY example we want to cycle the phase of the third pulse to select the pathways $\Delta p = -3$ and $+1$ and suppress all intermediate ones ($\Delta p = -2, -1$

Table 6.2. Possible changes in coherence order Δp without (A) and with (B) phase cycling of the final pulse in the DQF-COSY experiment.

A	B
+7	
+6	
+5	+5
+4	
+3	
+2	
+1	+1
0	
−1	
−2	
−3	−3
−4	
−5	
−6	
−7	−7
−8	
−9	

and 0). According to the above rule, to do this we must have $N = 4$. Deleting the pathways suppressed by the phase cycle we obtain column B in Table 6.2. Thus, our phase cycle also retains the pathways $\Delta p = -7$ and $+5$, i.e. it selects six- as well as two-quantum coherences during the delay. However, these will make only vanishing contributions to the final signal; even if the spin system is large enough to support such a high-order coherence, its excitation will be very inefficient.

In fact, no further phase cycling is necessary for DQF-COSY. Although there could be some z magnetization present during t_1 (either because the first pulse is not exactly 90°, or because of spin-lattice relaxation) it will not interfere because a single 90° pulse cannot convert z magnetization into multiple-quantum coherence. However, in more complicated experiments, selection of the desired pathway(s) will usually involve several independent phase cycles, which must be *nested* like loops in a FORTRAN or C computer program. Clearly an experiment with several cycles within cycles can end up being quite time-consuming; this is where pulsed field gradients come in.

6.4 Pulsed field gradients

The idea behind the use of field gradients for coherence pathway selection is that:

If a static magnetic field is applied briefly along the z axis, the phase shift experienced by a coherence will be proportional to its order, p.

The similarity to the First Rule of Phase Cycling should be clear. To be more specific, if a field of strength, B_G, is switched on for a time, τ, the phase shift (in radians) is $p\gamma B_G\tau$ where γ is the gyromagnetic ratio of the nucleus in question. If the field varies in strength across the NMR sample (which is what is meant by a field *gradient*), then coherences in different regions of the NMR tube experience different phase shifts. If the field gradient is strong enough and lasts long enough, coherences with non-zero orders will be completely dephased. A second pulsed field gradient, applied later in the pulse sequence, with suitable B_G and τ, can selectively undo the effect of the first, and so pick out the required change in coherence order. An example should clarify how this works.

Suppose we want to select the pathway $+2 \rightarrow -1$ at the end of a DQF-COSY sequence. If we put two pulsed field gradient pulses, labelled G_1 and G_2, either side of the 90° pulse, as in Fig. 6.2a, the total phase shift of the desired pathway will be

$$\phi = +2\gamma B_{G_1}\tau_1 - \gamma B_{G_2}\tau_2. \tag{6.6}$$

By making the second gradient twice as strong or twice as long as the first, we ensure that only the required pathway is rephased. Other pathways suffer non-zero net phase shifts and so remain dephased. Similarly, if the $-2 \rightarrow -1$ pathway is required, all that is needed is to reverse the direction of one of the gradients, as shown in Fig. 6.2b. Evidently, one cannot select *both* pathways simultaneously, in contrast to phase cycling. As a consequence, there is often

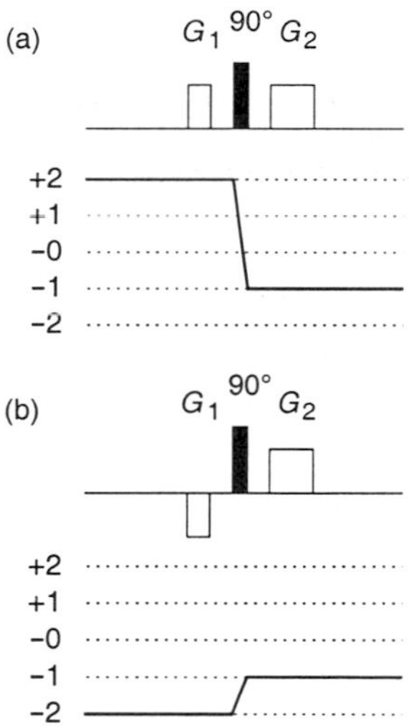

Fig. 6.2 Use of pulsed field gradients to select the different coherence transfer pathways: (a) $p = +2 \rightarrow p = -1$, and (b) $p = -2 \rightarrow p = -1$.

a signal-to-noise penalty associated with the use of field gradients. Coherence transfer pathways in heteronuclear experiments may be selected in exactly the same way, except that the appropriate gyromagnetic ratios must be included in expressions analogous to Eqn 6.6.

A 'gradient-selected' DQF-COSY pulse sequence and its coherence transfer pathway diagram are shown in Fig. 6.3. As described above, the final pulse is bracketed by two pulsed field gradients, $G_2 = -2G_1$ (i.e. $B_{G_2}\tau_2 = -2B_{G_1}\tau_1$). Acquisition starts immediately after G_2. Note that by not putting a gradient pulse into the t_1 period, it is still possible to use the method outlined in Section 2.6 (or similar) to get pure absorption lineshapes with sign discrimination in F_1.

In practice, extra steps need to be taken to refocus the phase errors that arise from evolution during Δ.

Pulsed field gradients have advantages over phase cycling that more than compensate for the modest loss in sensitivity noted above. Because the desired coherence transfer pathway can usually be selected in a single experiment, pulsed field gradients put a much smaller demand on the stability of the spectrometer than does phase cycling, where the unwanted signals are only cancelled at the end of a lengthy set of repetitions of the pulse sequence. Furthermore, field gradients allow one to arrange the experiment so as to obtain the desired resolution and sensitivity, without being constrained to complete an prolonged phase cycle for each t_1 value in a two-dimensional experiment.

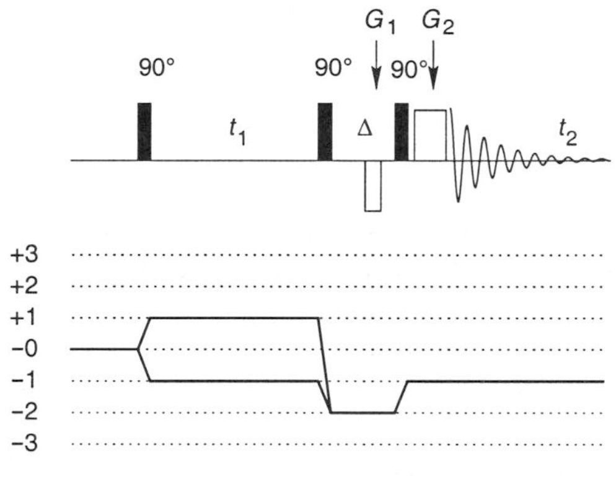

Fig. 6.3 DQF-COSY pulse sequence with pulsed field gradients instead of phase cycling.

7 Quantum mechanics

7.1 Introduction

In Part B of this book we turn to a rigorous quantum-mechanical description of NMR experiments. We explain what product operators really are, justify the way they have been used to understand the operation of NMR pulse sequences and derive many of the results that were quoted without proof in Part A. We begin with the quantum-mechanical description of a two-level system: this may seem very abstract, but we soon come to more concrete examples.

Throughout the next four chapters we assume a passing familiarity with basic quantum mechanics. Excellent textbooks include *Molecular Quantum Mechanics* by P. W. Atkins (we refer to this book as MQM) and the two Oxford Chemistry Primers by N. J. B. Green (OCP 48 and OCP 65).

7.2 Ket and bra vectors

The most convenient notation for describing two-level systems, such as spin-1/2 nuclei in magnetic fields, is the bra and ket notation developed by Dirac. Any arbitrary state of a two-level system is specified by a *ket* $|\psi\rangle$, which can in general be written as a linear combination, or *superposition*, of two basis states, one for each level:

See OCP 48 for a brief introduction to Dirac's bra(c)ket notation.

$$|\psi\rangle = c_1|1\rangle + c_2|2\rangle . \tag{7.1}$$

The basis states $|1\rangle$ and $|2\rangle$ can be thought of as orthogonal unit vectors, while the coefficients c_1 and c_2 measure the contributions of these two vectors to the superposition. Unlike conventional vectors, however, c_1 and c_2 can be complex numbers, and so do not simply describe the relative amounts of $|1\rangle$ and $|2\rangle$ in the superposition but also the relative phase with which they add together.

Following this vector approach, it is often convenient to write a ket as a column vector, whose elements are the coefficients in Eqn 7.1

$$\boldsymbol{\psi} = \begin{pmatrix} c_1 \\ c_2 \end{pmatrix} . \tag{7.2}$$

As before, vectors such as $\boldsymbol{\psi}$, are set in bold Roman type. We will switch back and forth between the $|\psi\rangle$ and $\boldsymbol{\psi}$ notations, using whichever is more convenient for our purposes.

Just as for conventional vectors we can define a scalar product, but it is necessary to allow for the fact that c_1 and c_2 can be complex. For each ket there is a corresponding *bra* $\langle\psi|$ defined by

Some important properties of vectors and matrices are summarized in Appendix 7.1.

$$\langle\psi| = \langle 1|c_1^* + \langle 2|c_2^* . \tag{7.3}$$

The asterisk indicates the complex conjugate.

This can be written as a row vector

$$\boldsymbol{\psi}^\dagger = \begin{pmatrix} c_1^* & c_2^* \end{pmatrix} \tag{7.4}$$

where † indicates the matrix adjoint (the complex conjugate of the matrix transpose). The scalar product of $\langle\psi|$ and $|\psi\rangle$ is then given by

Because the coefficients c_1 and c_2 are simply numbers they can be moved within this equation at will.

$$\begin{aligned}\langle\psi|\psi\rangle &= \left[\langle 1|c_1^* + \langle 2|c_2^*\right]\left[c_1|1\rangle + c_2|2\rangle\right] \\ &= c_1^*c_1\langle 1|1\rangle + c_1^*c_2\langle 1|2\rangle + c_2^*c_1\langle 2|1\rangle + c_2^*c_2\langle 2|2\rangle.\end{aligned} \tag{7.5}$$

Orthogonal unit vectors are said to be *orthonormal*.

Since $|1\rangle$ and $|2\rangle$ are orthogonal unit vectors, their scalar products are

$$\langle 1|1\rangle = \langle 2|2\rangle = 1 \qquad \langle 1|2\rangle = \langle 2|1\rangle = 0 \tag{7.6}$$

and so

The use of complex conjugates in defining the bra vector ensures that the lengths of vectors will be positive real numbers.

$$\langle\psi|\psi\rangle = c_1^*c_1 + c_2^*c_2 = |c_1|^2 + |c_2|^2, \tag{7.7}$$

which is the square of the length of $|\psi\rangle$. For convenience, $|\psi\rangle$ is usually *normalized* so that its length is equal to one.

This result can also be derived directly from the vector representations in Eqns 7.2 and 7.4: the scalar product is then just a conventional matrix product

$$\langle\psi|\psi\rangle = \boldsymbol{\psi}^\dagger\boldsymbol{\psi} = \begin{pmatrix} c_1^* & c_2^* \end{pmatrix}\begin{pmatrix} c_1 \\ c_2 \end{pmatrix} = c_1^*c_1 + c_2^*c_2\,. \tag{7.8}$$

7.3 Operators

Operators are described in more detail in Appendix 7.2.

An *operator* $\hat{A}$ acts on a ket $|\psi\rangle$ to give some new ket $|\psi'\rangle$. In general, the effect of $\hat{A}$ on $|\psi\rangle$ can be described as some combination of rotation and scaling. In some cases, however, $\hat{A}$ will simply scale (i.e. alter the length of) $|\psi\rangle$ without affecting its direction:

$$\hat{A}|\psi\rangle = a|\psi\rangle\,. \tag{7.9}$$

In this case $|\psi\rangle$ is said to be an *eigenvector* of $\hat{A}$, with *eigenvalue a*.

If the operator $\hat{A}$ corresponds to an observable quantity, such as energy or angular momentum, then the mean value obtained from an experimental measurement of this quantity is given by the *expectation value* of $\hat{A}$, which is simply the scalar product of $\langle\psi|$ with $\hat{A}|\psi\rangle$:

This expression assumes that $|\psi\rangle$ is normalized, so that $\langle\psi|\psi\rangle = 1$.

$$\langle\hat{A}\rangle = \langle\psi|\hat{A}|\psi\rangle\,. \tag{7.10}$$

Suppose that the basis vectors, $|1\rangle$ and $|2\rangle$, are eigenvectors of $\hat{A}$, that is

$$\hat{A}|1\rangle = a_1|1\rangle, \qquad \hat{A}|2\rangle = a_2|2\rangle\,. \tag{7.11}$$

The effect of $\hat{A}$ on $|\psi\rangle$ is then

$$\hat{A}|\psi\rangle = \hat{A}c_1|1\rangle + \hat{A}c_2|2\rangle = a_1c_1|1\rangle + a_2c_2|2\rangle \tag{7.12}$$

and so

$$\langle \hat{A} \rangle = \langle \psi | \hat{A} | \psi \rangle = \left[c_1^* \langle 1| + c_2^* \langle 2| \right] \left[a_1 c_1 |1\rangle + a_2 c_2 |2\rangle \right]$$
$$= a_1 |c_1|^2 + a_2 |c_2|^2. \tag{7.13}$$

Compare this expression with Eqns 7.5 and 7.7.

The mean value obtained from an experiment is thus a weighted average of the two eigenvalues, a_1 and a_2. Any single experiment will result in one of these two values being measured, with probabilities $|c_1|^2$ and $|c_2|^2$; these are the probabilities that the system will be found in the corresponding basis states.

In general, $|1\rangle$ and $|2\rangle$ are not eigenstates of $\hat{A}$, in which case

$$\langle \hat{A} \rangle = \left[\langle 1| c_1^* + \langle 2| c_2^* \right] \hat{A} \left[c_1 |1\rangle + c_2 |2\rangle \right]$$
$$= c_1^* c_1 \langle 1|\hat{A}|1\rangle + c_1^* c_2 \langle 1|\hat{A}|2\rangle + c_2^* c_1 \langle 2|\hat{A}|1\rangle + c_2^* c_2 \langle 2|\hat{A}|2\rangle. \tag{7.14}$$

It is useful to consider $\langle r|\hat{A}|s\rangle$ as an element A_{rs} of a matrix $\mathbf{A}$:

As for vectors, the symbols for matrices are set in bold roman type.

$$\mathbf{A} = \begin{pmatrix} A_{11} & A_{12} \\ A_{21} & A_{22} \end{pmatrix} = \begin{pmatrix} \langle 1|\hat{A}|1\rangle & \langle 1|\hat{A}|2\rangle \\ \langle 2|\hat{A}|1\rangle & \langle 2|\hat{A}|2\rangle \end{pmatrix}. \tag{7.15}$$

This allows Eqn 7.14 to be written as

$$\langle \hat{A} \rangle = c_1^* c_1 A_{11} + c_1^* c_2 A_{12} + c_2^* c_1 A_{21} + c_2^* c_2 A_{22}, \tag{7.16}$$

or, much more compactly, using the vector notation introduced above:

$$\langle \hat{A} \rangle = \begin{pmatrix} c_1^* & c_2^* \end{pmatrix} \begin{pmatrix} A_{11} & A_{12} \\ A_{21} & A_{22} \end{pmatrix} \begin{pmatrix} c_1 \\ c_2 \end{pmatrix} = \psi^\dagger \mathbf{A} \psi . \tag{7.17}$$

Compare this expression with Eqns 7.8 and 7.10.

7.4 Angular momentum

We now turn from these general expressions to a more concrete example. One of the most important operators in the quantum-mechanical description of any system is the *Hamiltonian*, which determines the total energy of the system,

$$E = \langle \hat{H} \rangle = \langle \psi | \hat{H} | \psi \rangle . \tag{7.18}$$

As we shall see, the Hamiltonians which occur in NMR are closely related to the operators which describe angular momentum; these operators play a central role in the theory of NMR.

The energy of a classical magnetic dipole, such as a compass needle, in a magnetic field is given by $E = -\boldsymbol{\mu}\cdot\mathbf{B_0} = -\mu_z B_0$, where $\boldsymbol{\mu}$ is the magnetic moment, μ_z is its z component, and the magnetic field, $\mathbf{B_0}$, is along the z axis. For a particle such as an atomic nucleus, the magnetic moment is proportional to its angular momentum, so that $\mu_z = \gamma I_z$, where the proportionality coefficient γ is the *gyromagnetic ratio*. The analogous quantum-mechanical

See OCP 32 for a discussion of this expression.

expression can be obtained by replacing E and I_z by the corresponding quantum mechanical operators:

$$\hat{H} = -\gamma B_0 \hbar \hat{I}_z = \hbar\omega_0 \hat{I}_z \tag{7.19}$$

Writing energies as multiples of $\hbar$ (Planck's constant over 2π) or, equivalently, working with units such that $\hbar = 1$ is a common and convenient practice in NMR.

where $\omega_0 = -\gamma B_0$ is an angular frequency (the *Larmor* frequency). For simplicity, energies may be written as multiples of $\hbar$, so that

$$\hat{H} = \omega_0 \hat{I}_z\,. \tag{7.20}$$

The eigenstates of $\hat{I}_z$, and thus of the Hamiltonian, are conventionally called $|\alpha\rangle$ (spin-up) and $|\beta\rangle$ (spin-down); these states are a natural choice for basis states. The corresponding eigenvalues are given by

$$\hat{I}_z|\alpha\rangle = +\tfrac{1}{2}|\alpha\rangle \qquad \hat{I}_z|\beta\rangle = -\tfrac{1}{2}|\beta\rangle \tag{7.21}$$

$$\hat{H}|\alpha\rangle = +\tfrac{1}{2}\omega_0|\alpha\rangle \qquad \hat{H}|\beta\rangle = -\tfrac{1}{2}\omega_0|\beta\rangle, \tag{7.22}$$

hence $|\alpha\rangle$ and $|\beta\rangle$ are separated in energy by ω_0. For many nuclei, including ^1H, γ is positive and so ω_0 is negative; thus $|\alpha\rangle$ has a lower energy than $|\beta\rangle$.

Matrix representations are calculated as shown in Eqn 7.15. N.B. $|\alpha\rangle$ and $|\beta\rangle$ are orthonormal.

The eigenvalues in Eqn 7.21 can be used to determine the *matrix representation* of $\hat{I}_z$:

$$\mathbf{I_z} = \begin{pmatrix} \frac{1}{2} & 0 \\ 0 & -\frac{1}{2} \end{pmatrix}. \tag{7.23}$$

From Eqn 7.13, the expectation value is

$$\langle \hat{I}_z \rangle = \tfrac{1}{2}|c_\alpha|^2 - \tfrac{1}{2}|c_\beta|^2, \tag{7.24}$$

that is, half the difference between the probabilities of being found in the two states.

Unlike $\hat{I}_z$, the operators corresponding to x and y angular momentum do not have $|\alpha\rangle$ and $|\beta\rangle$ as eigenstates. Instead, these operators interconvert $|\alpha\rangle$ and $|\beta\rangle$:

Justifying Eqns 7.21 and 7.25 is not a simple business. One approach is to note that these forms give the correct commutation relationships, as shown below (Eqn 7.30).

$$\begin{aligned} \hat{I}_x|\alpha\rangle &= \tfrac{1}{2}|\beta\rangle \qquad \hat{I}_x|\beta\rangle = \tfrac{1}{2}|\alpha\rangle \\ \hat{I}_y|\alpha\rangle &= \tfrac{1}{2}i|\beta\rangle \qquad \hat{I}_y|\beta\rangle = -\tfrac{1}{2}i|\alpha\rangle. \end{aligned} \tag{7.25}$$

As before, these expressions can be expressed more compactly in matrix form

The matrices in Eqns 7.23 and 7.26 are equivalent to the Pauli matrices. See MQM for a more detailed discussion.

$$\mathbf{I_x} = \begin{pmatrix} 0 & \frac{1}{2} \\ \frac{1}{2} & 0 \end{pmatrix} \qquad \mathbf{I_y} = \begin{pmatrix} 0 & -\frac{1}{2}i \\ \frac{1}{2}i & 0 \end{pmatrix}. \tag{7.26}$$

From Eqn 7.16, their expectation values are

Re(a) and Im(a) indicate the real and imaginary parts of a.

$$\begin{aligned} \langle \hat{I}_x \rangle &= \tfrac{1}{2}\left(c_\alpha c_\beta^* + c_\alpha^* c_\beta\right) = \mathrm{Re}\left(c_\alpha c_\beta^*\right) \\ \langle \hat{I}_y \rangle &= \tfrac{1}{2}i\left(c_\alpha c_\beta^* - c_\alpha^* c_\beta\right) = -\mathrm{Im}\left(c_\alpha c_\beta^*\right). \end{aligned} \tag{7.27}$$

Note that transverse magnetization depends on the *cross-products* $c_\alpha c_\beta^*$ and $c_\alpha^* c_\beta$, while the longitudinal component z is determined by the *self-products* $c_\alpha c_\alpha^*$ and $c_\beta c_\beta^*$.

The matrix representations, Eqns 7.23 and 7.26, can be used to deduce further properties of the angular momentum operators:

$$\hat{I}_x^2 = \hat{I}_x \hat{I}_x = \tfrac{1}{4}\hat{1}, \qquad (7.28)$$

(and similarly for $\hat{I}_y$ and $\hat{I}_z$), where $\hat{1}$ is the unit operator;

$$\hat{I}_x \hat{I}_y = -\hat{I}_y \hat{I}_x = \tfrac{1}{2} i \hat{I}_z \qquad (7.29)$$

$$\left[\hat{I}_x, \hat{I}_y\right] \equiv \hat{I}_x \hat{I}_y - \hat{I}_y \hat{I}_x = i \hat{I}_z \qquad (7.30)$$

(and equivalent expressions obtained by cyclic permutation of the subscripts, $x \to y \to z \to x$). The expression in Eqn 7.30, the *operator commutator*, will be of great importance later on.

These *commutation relations* must hold for any set of operators which describe angular momentum; effectively they can be taken as *defining* the forms of the operators. They can themselves be deduced from the general properties of angular momentum and the fundamental position–momentum commutation relationship, as described in MQM.

7.5 Free precession

Now we have the tools needed to look at the NMR behaviour of a single isolated spin-1/2 particle. First, we consider the time-dependence of the system during a period of free precession. This may be obtained from the *time-dependent Schrödinger equation*

$$\frac{\mathrm{d}}{\mathrm{d}t}|\psi\rangle = -i\hat{H}|\psi\rangle \qquad (7.31)$$

with the general initial state $|\psi\rangle = c_\alpha |\alpha\rangle + c_\beta |\beta\rangle$. Thus

$$\frac{\mathrm{d}}{\mathrm{d}t}|\psi\rangle = \frac{\mathrm{d}c_\alpha}{\mathrm{d}t}|\alpha\rangle + \frac{\mathrm{d}c_\beta}{\mathrm{d}t}|\beta\rangle = -ic_\alpha \hat{H}|\alpha\rangle - ic_\beta \hat{H}|\beta\rangle . \qquad (7.32)$$

Note that the basis states cannot vary with time, and so the time dependence of a state $|\psi\rangle$ must result from variations in its coefficients.

Multiplying from the left by $\langle\alpha|$ gives

$$\frac{\mathrm{d}c_\alpha}{\mathrm{d}t}\langle\alpha|\alpha\rangle + \frac{\mathrm{d}c_\beta}{\mathrm{d}t}\langle\alpha|\beta\rangle = -ic_\alpha\langle\alpha|\hat{H}|\alpha\rangle - ic_\beta\langle\alpha|\hat{H}|\beta\rangle \qquad (7.33)$$

or, using the fact that $|\alpha\rangle$ and $|\beta\rangle$ are orthonormal,

$$\frac{\mathrm{d}c_\alpha}{\mathrm{d}t} = -ic_\alpha H_{\alpha\alpha} - ic_\beta H_{\alpha\beta} . \qquad (7.34)$$

Similarly, multiplying Eqn 7.32 from the left by $\langle\beta|$ gives a corresponding expression for $\mathrm{d}c_\beta/\mathrm{d}t$. These expressions can be expressed more concisely in matrix form as

$$\begin{pmatrix} \dot{c}_\alpha \\ \dot{c}_\beta \end{pmatrix} = -i \begin{pmatrix} H_{\alpha\alpha} & H_{\alpha\beta} \\ H_{\beta\alpha} & H_{\beta\beta} \end{pmatrix} \begin{pmatrix} c_\alpha \\ c_\beta \end{pmatrix} \qquad (7.35)$$

The dots indicate differentiation with respect to time.

or

$$\dot{\psi} = -i\mathbf{H}\psi\,. \tag{7.36}$$

The form of this equation is identical to that of the conventional time-dependent Schrödinger equation (Eqn 7.31).

Using Eqns 7.20 and 7.23

$$\mathbf{H} = \omega_0 \mathbf{I_z} = \begin{pmatrix} \frac{1}{2}\omega_0 & 0 \\ 0 & -\frac{1}{2}\omega_0 \end{pmatrix} \tag{7.37}$$

so that Eqn 7.35 gives

$$\dot{c}_\alpha = -\tfrac{1}{2} i\omega_0 c_\alpha, \qquad \dot{c}_\beta = \tfrac{1}{2} i\omega_0 c_\beta. \tag{7.38}$$

Integration of these expressions gives

$$c_\alpha(t) = c_\alpha(0)\mathrm{e}^{-\frac{1}{2}i\omega_0 t} \qquad c_\beta(t) = c_\beta(0)\mathrm{e}^{\frac{1}{2}i\omega_0 t} \tag{7.39}$$

i.e. the two coefficients oscillate at frequencies $\pm\omega_0/2$.

Suppose we start from the initial condition $c_\alpha(0) = c_\beta(0) = 1/\sqrt{2}$, that is $\langle\hat{I}_x\rangle = 1/2$, $\langle\hat{I}_y\rangle = \langle\hat{I}_z\rangle = 0$; then from Eqns 7.24 and 7.27

$$\begin{aligned} \langle\hat{I}_x\rangle &= \mathrm{Re}\left[c_\alpha(0)c_\beta^*(0)\mathrm{e}^{-i\omega_0 t}\right] = \tfrac{1}{2}\cos\omega_0 t \\ \langle\hat{I}_y\rangle &= -\mathrm{Im}\left[c_\alpha(0)c_\beta^*(0)\mathrm{e}^{-i\omega_0 t}\right] = \tfrac{1}{2}\sin\omega_0 t \\ \langle\hat{I}_z\rangle &= \tfrac{1}{2}\left[c_\alpha(0)c_\alpha^*(0) - c_\beta(0)c_\beta^*(0)\right] = 0. \end{aligned} \tag{7.40}$$

This is *exactly* what is expected from the vector model for free precession in the laboratory frame (neglecting relaxation): the z magnetization does not change, while the x and y magnetizations oscillate at the Larmor frequency.

7.6 Radiofrequency pulses

Working in a rotating frame, as described later in Appendix 8.4, greatly simplifies things, and is almost essential for all but the simplest problems (see Section 1.3). Note the similarity between Eqns 7.41 and 7.20.

The effects of radiofrequency pulses can be treated in a similar way. At resonance, with the radiofrequency field B_1 taken along the x axis of the rotating frame,

$$\hat{H} = -\gamma B_1 \hat{I}_x = \omega_1 \hat{I}_x\,. \tag{7.41}$$

Proceeding in the same manner as above gives

$$\dot{c}_\alpha = -\tfrac{1}{2} i\omega_1 c_\beta \tag{7.42}$$

$$\dot{c}_\beta = -\tfrac{1}{2} i\omega_1 c_\alpha. \tag{7.43}$$

These coupled differential equations can be uncoupled by differentiating Eqn 7.42 once again

$$\ddot{c}_\alpha = \frac{\mathrm{d}}{\mathrm{d}t}\dot{c}_\alpha = -\tfrac{1}{2}i\omega_1\dot{c}_\beta = -\tfrac{1}{4}\omega_1^2 c_\alpha . \tag{7.44}$$

The general solution of this differential equation is $c_\alpha(t) = A\cos(\omega_1 t/2) + B\sin(\omega_1 t/2)$ where A and B are constants determined by the boundary conditions.

Starting from the spin-up state, $|\alpha\rangle$, i.e. $c_\alpha(0) = 1$ and $c_\beta(0) = 0$, Eqn 7.44 gives

$$c_\alpha(t) = \cos\left(\tfrac{1}{2}\omega_1 t\right). \tag{7.45}$$

Substituting this result back into Eqn 7.43 and integrating gives

$$c_\beta(t) = -i\sin\left(\tfrac{1}{2}\omega_1 t\right). \tag{7.46}$$

Hence

$$\begin{aligned}
\langle \hat{I}_x \rangle &= \mathrm{Re}\left[\cos\left(\tfrac{1}{2}\omega_1 t\right) i \sin\left(\tfrac{1}{2}\omega_1 t\right)\right] = 0 \\
\langle \hat{I}_y \rangle &= -\mathrm{Im}\left[\cos\left(\tfrac{1}{2}\omega_1 t\right) i \sin\left(\tfrac{1}{2}\omega_1 t\right)\right] = -\tfrac{1}{2}\sin\omega_1 t \\
\langle \hat{I}_z \rangle &= \tfrac{1}{2}\left[\cos^2\left(\tfrac{1}{2}\omega_1 t\right) - \sin^2\left(\tfrac{1}{2}\omega_1 t\right)\right] = \tfrac{1}{2}\cos\omega_1 t.
\end{aligned} \tag{7.47}$$

Thus, magnetization starting along the z axis rotates at a frequency ω_1 around the x axis towards the negative y axis (cf. Sections 1.4 and 3.2), as expected.

In principle, this method could be extended to deal with more complex NMR experiments. There is, however, a simpler approach (Chapter 8).

Appendix 7.1. Vectors and matrices

A *matrix* is an array of numbers arranged in rows and columns, e.g.

$$\mathbf{M} = \begin{pmatrix} p & q \\ r & s \end{pmatrix}. \tag{7.48}$$

The individual numbers making up a matrix are called *elements*, and are addressed according to their position in the matrix by stating first their *row* number and then their *column* number; these are normally written as subscripts and so, for example, we write $M_{12} = q$. Note that the name of a matrix is printed in bold roman type, unlike the individual elements.

If a matrix has only a single row it is called a *row vector*; if it has only a single column it is called a *column vector*, or just a *vector*. For example

$$\mathbf{a} = \begin{pmatrix} a_x \\ a_y \\ a_z \end{pmatrix} \tag{7.49}$$

is a vector with three elements, the x, y, and z components of **a**.

Two matrices, **M** and **N**, may be added or multiplied as follows. The sum **S** = **M** + **N**, has elements given by

$$S_{rc} = M_{rc} + N_{rc} \tag{7.50}$$

while the product, **P** = **MN**, has

$$P_{rc} = \sum_j M_{rj} N_{jc}. \tag{7.51}$$

It is only possible to add two matrices if they have the same shape, that is the same number of rows and columns. Similarly, it is only possible to multiply two matrices if the first matrix has as many columns as the second has rows. Matrix addition is *commutative*, that is $\mathbf{M} + \mathbf{N} = \mathbf{N} + \mathbf{M}$, but matrix multiplication is not, and so $\mathbf{MN} \neq \mathbf{NM}$ in general. Sometimes, however, $\mathbf{MN}$ does equal $\mathbf{NM}$, and in such cases the two matrices $\mathbf{M}$ and $\mathbf{N}$ are said to *commute*. Clearly, two matrices can only commute if they are both *square* (i.e. have the same number of rows as columns) and of the same size; if this is not the case then at least one of the two matrix multiplications cannot be performed.

In addition to these *binary* matrix operations, which combine two matrices to produce a third, there are also *unary* operations, which act on a single matrix. One simple example is the matrix *transpose*, defined by interchanging rows and columns; for square matrices this corresponds to swapping the elements around the *principal diagonal*. For our example matrix,

$$\mathbf{M}^{\mathsf{T}} = \begin{pmatrix} p & r \\ q & s \end{pmatrix}. \tag{7.52}$$

Note that transposition will convert a column vector into a row vector and vice versa.

There is one more matrix operation, which we shall need early in Chapter 8: the *trace*, which is the sum of the elements on the principal diagonal of a square matrix. A useful property of the trace of a *product* of matrices is that it does not change if the matrices are *cyclically permuted*, e.g. $\mathbf{ABC} \to \mathbf{CAB} \to \mathbf{BCA}$. This may be demonstrated easily:

$$\begin{aligned} \mathrm{Tr}[\mathbf{ABC}] &= \sum_j (\mathbf{ABC})_{jj} = \sum_j \sum_k \sum_m A_{jk} B_{km} C_{mj} \\ &= \sum_j \sum_k \sum_m C_{mj} A_{jk} B_{km} = \sum_m (\mathbf{CAB})_{mm} = \mathrm{Tr}[\mathbf{CAB}] \\ &= \sum_j \sum_k \sum_m B_{km} C_{mj} A_{jk} = \sum_k (\mathbf{BCA})_{kk} = \mathrm{Tr}[\mathbf{BCA}]. \end{aligned} \tag{7.53}$$

A_{jk}, B_{km}, and C_{mj} are just numbers, and can be multiplied in any order.

The result holds regardless of how many matrices are multiplied within the trace, and however many are cyclically permuted.

Turning now to vectors, it is clear that one can add two vectors by adding corresponding elements, as long as the vectors have the same size; this is just a special case of matrix addition. Similarly, two vectors can be multiplied to form the *vector dot product* or *scalar product* by multiplying corresponding elements and adding them together. For two vectors, **a** and **b**,

$$\mathbf{a} \cdot \mathbf{b} = a_x b_x + a_y b_y + a_z b_z. \tag{7.54}$$

The dot product can also be viewed as a special case of matrix multiplication,

$$\mathbf{a} \cdot \mathbf{b} = \mathbf{a}^{\mathsf{T}} \mathbf{b}, \tag{7.55}$$

but it is sufficiently important to be considered as an operation in its own right. For example, the dot product can be used to determine the length, a, of a vector **a**:

$$a = \sqrt{a_x^2 + a_y^2 + a_z^2} = \sqrt{\mathbf{a} \cdot \mathbf{a}} \, . \qquad (7.56)$$

Similarly for two vectors, **a** and **b**,

$$\mathbf{a} \cdot \mathbf{b} = ab \cos\theta \, , \qquad (7.57)$$

where θ is the angle between the two vectors. If $\theta = 90°$ then $\mathbf{a} \cdot \mathbf{b} = 0$, and the two vectors are said to be *orthogonal*.

Appendix 7.2. Operators

Operators are mathematical devices which take in a mathematical object of some kind and convert it to another (usually different) object of the same kind. In many elementary treatments of quantum mechanics for chemists, operators are depicted as acting on functions; for example, differentiation is an operator acting on a function to give another function.

Within the Dirac bra and ket formalism, operators act on kets to produce other kets. As these kets may be thought of as complex vectors, operators can be described using complex square matrices. In many cases it is simplest to discuss the properties of operators in terms of the properties of the corresponding matrices.

The product of two operators, $\hat{A}_1\hat{A}_2$, is equivalent to operating first with the operator $\hat{A}_2$ and then with the operator $\hat{A}_1$. More simply, the matrix corresponding to $\hat{A}_1\hat{A}_2$ is the matrix product $\mathbf{A_1A_2}$. Just as the order in which matrices are multiplied is important (generally, $\mathbf{A_1A_2} \neq \mathbf{A_2A_1}$), the effect of two operators will usually depend on the order in which they are applied. This effect is summarized by the *operator commutator*,

$$\left[\hat{A}_1, \hat{A}_2\right] = \hat{A}_1\hat{A}_2 - \hat{A}_2\hat{A}_1 \, . \qquad (7.58)$$

If the operator commutator is zero, then the operators are said to *commute*, and it does not matter in which order they are applied.

The *inverse* of an operator $\hat{A}$ is written as $\hat{A}^{-1}$, and is defined such that

$$\hat{A}\hat{A}^{-1} = \hat{A}^{-1}\hat{A} = \hat{1} \qquad (7.59)$$

where $\hat{1}$ is the unit or identity operator, which has no effect on any ket; the corresponding matrix is the identity matrix, **1**, for which all the elements are zero except those on the principal diagonal, which are all equal to one. While it may be difficult to determine an operator inverse, the matrix inverse is easily found using numerical techniques. Another important concept, which is most easily defined in terms of matrices, is the operator *adjoint*: the adjoint of a matrix **A**, denoted $\mathbf{A}^\dagger$, is obtained by taking the complex conjugate of the matrix transpose.

The adjoint and inverse can be used to define two important groups of matrices, and hence operators. A matrix is said to be *Hermitian*, or self-adjoint, if $\mathbf{A} = \mathbf{A}^\dagger$, while it is *unitary* if $\mathbf{A}^{-1} = \mathbf{A}^\dagger$. Operators that correspond to *observables* (such as the Hamiltonian, which corresponds to the total energy) are always Hermitian.

Another important property is the existence of *eigenvectors* of operators; these have the property that they are scaled by the operator, but otherwise left unchanged. Eigenvectors and their corresponding *eigenvalues* will be explored in more detail in Appendix 8.2.

8 Density matrices

8.1 Introduction

The methods outlined in the previous chapter are adequate for simple problems, but swiftly become tedious for more complicated cases involving two or more spins. Fortunately, there is a more compact and convenient approach.

Density matrices are described in many texts on quantum mechanics: excellent treatments with an emphasis on NMR can be found in the books of Munowitz (1988) and Goldman (1988).

8.2 The density operator

As shown in the previous chapter, expectation values of an operator $\hat{A}$ depend on *products* of coefficients. For an operator acting on a single spin (Eqn 7.16)

$$\langle \hat{A} \rangle = c_\alpha^* c_\alpha A_{\alpha\alpha} + c_\alpha^* c_\beta A_{\alpha\beta} + c_\beta^* c_\alpha A_{\beta\alpha} + c_\beta^* c_\beta A_{\beta\beta}. \tag{8.1}$$

It is therefore useful to define a *density operator* which is most conveniently described in terms of the elements of its corresponding *density matrix*, $\boldsymbol{\rho}$:

$$\rho_{rs} = \langle r|\hat{\rho}|s\rangle = c_r c_s^* . \tag{8.2}$$

Combining Eqns 8.1 and 8.2,

$$\langle \hat{A} \rangle = \rho_{\alpha\alpha} A_{\alpha\alpha} + \rho_{\alpha\beta} A_{\beta\alpha} + \rho_{\beta\alpha} A_{\alpha\beta} + \rho_{\beta\beta} A_{\beta\beta} \tag{8.3}$$

which is simply the *trace* of the product of $\boldsymbol{\rho}$ and $\mathbf{A}$:

$$\langle \hat{A} \rangle = \sum_j \sum_k \rho_{jk} A_{kj} = \sum_j (\boldsymbol{\rho}\mathbf{A})_{jj} = \mathrm{Tr}(\boldsymbol{\rho}\mathbf{A}). \tag{8.4}$$

The trace of a matrix is the sum of the elements on its principal diagonal (see Appendix 7.1).

Eqn 8.2 shows how the density matrix $\boldsymbol{\rho}$ can be calculated directly from the underlying coefficients. For a single spin

$$\boldsymbol{\rho} = \begin{pmatrix} c_\alpha \\ c_\beta \end{pmatrix} \begin{pmatrix} c_\alpha^* & c_\beta^* \end{pmatrix} = \begin{pmatrix} c_\alpha c_\alpha^* & c_\alpha c_\beta^* \\ c_\beta c_\alpha^* & c_\beta c_\beta^* \end{pmatrix} \tag{8.5}$$

Note that the product of a column vector and a row vector is a square *matrix*, while the product of a row vector and a column vector is a *number*.

or, in general,

$$\boldsymbol{\rho} = \boldsymbol{\psi}\boldsymbol{\psi}^\dagger . \tag{8.6}$$

Using the analogy between vectors and kets, the density operator can be written in terms of bras and kets:

$$\hat{\rho} = |\psi\rangle\langle\psi| . \tag{8.7}$$

See also Section 8.4 and Appendix 8.1.

In Chapter 7 we saw that the expectation value of $\hat{I}_z$ is determined by $c_\alpha c_\alpha{}^*$ and $c_\beta c_\beta{}^*$ while that of $\hat{I}_x$ and $\hat{I}_y$ depends on $c_\alpha c_\beta{}^*$ and $c_\beta c_\alpha{}^*$. Thus, the diagonal elements of the density matrix represent *populations* while the off-diagonal elements represent *coherences* (in this case, single-quantum coherences).

Next, we need to determine how ρ evolves under a given Hamiltonian. It is sufficient to calculate the behaviour of a single element:

This expression is obtained from Eqn 8.2 by the standard rules for differentiating a product. As in the previous chapter the dot indicates differentiation with respect to time.

$$\dot{\rho}_{rs} = \dot{c}_r c_s^* + c_r \dot{c}_s^* . \tag{8.8}$$

Generalizing from Eqn 7.35,

$$\begin{aligned} \dot{c}_r &= -i\sum_k H_{rk} c_k \\ \dot{c}_s^* &= +i\sum_k H_{sk}^* c_k^* \end{aligned} \tag{8.9}$$

and so

$$\begin{aligned} \dot{\rho}_{rs} &= -i\sum_k H_{rk} c_k c_s^* + i\sum_k c_r c_k^* H_{sk}^* \\ &= -i\sum_k H_{rk} \rho_{ks} + i\sum_k \rho_{rk} H_{ks} \\ &= -i(\mathbf{H}\boldsymbol{\rho})_{rs} + i(\boldsymbol{\rho}\mathbf{H})_{rs} = -i[\mathbf{H},\boldsymbol{\rho}]_{rs} \end{aligned} \tag{8.10}$$

See Appendices 7.1 and 7.2 for elementary properties of matrices and operators.

where we have used the fact that **H** is Hermitian (i.e. $\mathbf{H} = \mathbf{H}^\dagger$, or $H_{sk}{}^* = H_{ks}$). Thus, since Eqn 8.10 holds for each of the elements of ρ,

$$\frac{\mathrm{d}\hat{\rho}}{\mathrm{d}t} = -i\left[\hat{H},\hat{\rho}\right]. \tag{8.11}$$

This equation, which relates the evolution of the density operator to the Hamiltonian, plays a central role in the theory of density matrices. It is important enough to have a name: the *Liouville–von Neumann equation*.

8.3 Solving the Liouville–von Neumann equation

The Liouville–von Neumann equation is readily solved when the Hamiltonian is constant for some time, *t*. The solution is most easily written in terms of *operator exponentials*

$$\hat{\rho}(t) = \mathrm{e}^{-i\hat{H}t}\hat{\rho}(0)\mathrm{e}^{+i\hat{H}t} \tag{8.12}$$

where

More practical approaches for calculating operator exponentials are described in Appendix 8.2.

$$\mathrm{e}^{\hat{A}} = \hat{1} + \hat{A} + \frac{\hat{A}^2}{2!} + \frac{\hat{A}^3}{3!} + \cdots \tag{8.13}$$

and $\hat{A}^n$ is a convenient shorthand for applying $\hat{A}$ *n* times. This solution can be verified by differentiating Eqn 8.12:

$$\frac{d\hat{\rho}}{dt} = \left(\frac{de^{-i\hat{H}t}}{dt}\right)\hat{\rho}(0)e^{+i\hat{H}t} + e^{-i\hat{H}t}\hat{\rho}(0)\left(\frac{de^{+i\hat{H}t}}{dt}\right)$$
$$= -i\hat{H}e^{-i\hat{H}t}\hat{\rho}(0)e^{+i\hat{H}t} + e^{-i\hat{H}t}\hat{\rho}(0)e^{+i\hat{H}t}i\hat{H} \quad (8.14)$$
$$= -i\hat{H}\hat{\rho} + i\hat{\rho}\hat{H} = -i\left[\hat{H},\hat{\rho}\right],$$

The initial density operator is, of course, independent of t.

where we have used the fact that

$$\frac{de^{k\hat{A}t}}{dt} = k\hat{A}\,e^{k\hat{A}t} = e^{k\hat{A}t}k\hat{A} \quad (8.15)$$

This expression may be derived from the exponential expansion (Eqn 8.13) by analogy with more conventional exponentials. Note that any operator commutes with any power of itself and therefore with its operator exponential.

for any real or complex number k.

Solving the Liouville–von Neumann equation is more difficult when the Hamiltonian varies with time. If, however, the Hamiltonian is constant during individual periods making up the entire evolution time, then the equation is easily solved by an extension of Eqn 8.12:

$$\hat{\rho}(0) \xrightarrow{\hat{H}_1 t_1} e^{-i\hat{H}_1 t_1}\hat{\rho}(0)e^{+i\hat{H}_1 t_1}$$
$$\xrightarrow{\hat{H}_2 t_2} e^{-i\hat{H}_2 t_2}e^{-i\hat{H}_1 t_1}\hat{\rho}(0)e^{+i\hat{H}_1 t_1}e^{+i\hat{H}_2 t_2} \quad (8.16)$$

and so on.

8.4 Ensemble averages

The density operator is certainly a convenient mathematical device, which avoids explicitly calculating coefficients, such as c_α, when it is only their products, such as $c_\alpha c_\beta{}^*$, which actually determine the values of observable quantities (e.g. Eqns 7.24 and 7.27). It has a second great advantage, however, in that it provides an elegant way of treating macroscopic systems.

Many macroscopic systems, such as NMR samples, are made up of a very large number of identical independent copies (effectively an ensemble of copies) of individual microscopic systems. Each microscopic system feels the same basic interactions and thus evolves under the same Hamiltonian, but the systems do not all evolve in the same way because each copy has its own initial state.

By an ensemble of independent copies we mean that the individual microscopic systems do not interact with one another in any way.

Consider, for example, two microscopic systems with initial states $|\psi_1\rangle$ and $|\psi_2\rangle$ respectively. Using the bra–ket approach outlined in the previous chapter it is necessary to evaluate the behaviour of each system individually: it is not possible simply to add the two kets together. The same problem occurs with density operators, but in this case a great simplification can be made as long as it is not necessary to determine individual expectation values for the two microscopic systems, but only their sum. This is a realistic restriction, since it corresponds exactly to performing experiments on a macroscopic system containing two or more indistinguishable microscopic systems. In this case

$$\left\langle\hat{A}\right\rangle_1 + \left\langle\hat{A}\right\rangle_2 = \mathrm{Tr}(\rho_1\mathbf{A}) + \mathrm{Tr}(\rho_2\mathbf{A}) = \mathrm{Tr}([\rho_1+\rho_2]\mathbf{A}), \quad (8.17)$$

so it is possible to add density operators before evaluating expectation values. Similarly,

$$\begin{aligned}\hat{\rho}_1(t)+\hat{\rho}_2(t) &= \mathrm{e}^{-i\hat{H}t}\hat{\rho}_1(0)\mathrm{e}^{+i\hat{H}t}+\mathrm{e}^{-i\hat{H}t}\hat{\rho}_2(0)\mathrm{e}^{+i\hat{H}t} \\ &= \mathrm{e}^{-i\hat{H}t}\left[\hat{\rho}_1(0)+\hat{\rho}_2(0)\right]\mathrm{e}^{+i\hat{H}t}.\end{aligned} \tag{8.18}$$

Thus, it is possible to define a summed density operator

$$\hat{\rho}(t)=\hat{\rho}_1(t)+\hat{\rho}_2(t) \tag{8.19}$$

which can be used in any further calculations.

Remember that this approach is only applicable when the microscopic systems are *identical* and *independent*. If this is not the case, it is necessary to use a larger density matrix that explicitly describes each microscopic system.

The same approach can be used with any number of microscopic systems: the corresponding density operators can be summed to produce an overall density operator. However, it is more usual to *average* the individual density operators. Thus, in general, the density operator of a macroscopic system is simply the ensemble average of the density operators of the underlying microscopic systems,

$$\hat{\rho}=\sum_i p_i\hat{\rho}_i\,, \tag{8.20}$$

where p_i is the probability of a microscopic system being in state $\hat{\rho}_i$. The individual elements of $\boldsymbol{\rho}$ are given by

$$\rho_{rs}=\overline{c_r c_s^*} \tag{8.21}$$

where the bar indicates an ensemble average.

Appendix 8.1 discusses in a little more detail the difference between *mixed* states such as this, and *pure* states where all the constituent microscopic systems are identical.

A particularly important example of an ensemble-averaged density operator is that of a macroscopic system at thermal equilibrium (the usual starting point for an NMR experiment). In a sample of molecules, each containing a single spin-1/2 nucleus, there is a mixture of spins in the states $|\alpha\rangle$ and $|\beta\rangle$; thus

Compare this expression with Eqn 8.7.

$$\hat{\rho}=p_\alpha|\alpha\rangle\langle\alpha|+p_\beta|\beta\rangle\langle\beta| \tag{8.22}$$

or

$$\boldsymbol{\rho}=\begin{pmatrix}p_\alpha & 0\\ 0 & p_\beta\end{pmatrix} \tag{8.23}$$

where p_α and p_β are the probabilities of spins being in the two states as given by the Boltzmann distribution. The density operator can be rewritten:

$$\begin{aligned}\boldsymbol{\rho} &= \tfrac{1}{2}\begin{pmatrix}p_\alpha+p_\beta & 0\\ 0 & p_\alpha+p_\beta\end{pmatrix}+\tfrac{1}{2}\begin{pmatrix}p_\alpha-p_\beta & 0\\ 0 & -p_\alpha+p_\beta\end{pmatrix} \\ &= \tfrac{1}{2}\mathbf{1}+\left(p_\alpha-p_\beta\right)\mathbf{I_z}.\end{aligned} \tag{8.24}$$

As the unit matrix, **1**, does not evolve under any Hamiltonian, all the interesting behaviour may be deduced by examining the evolution of the $\mathbf{I_z}$

term; $(p_\alpha - p_\beta)$ is just a scaling factor and is often omitted, e.g. in Eqn 8.27 below.

8.5 Application to NMR

To illustrate how density matrices can be used, we now calculate the effect of an on-resonance x pulse in the rotating frame, with the system initially in the mixed state $\hat{I}_z$. As before, the Hamiltonian is given by

Note that we use the same notation to describe both the state of the system and our operations upon it.

$$\hat{H} = \omega_1 \hat{I}_x \tag{8.25}$$

so that (using Eqn 8.12)

$$\rho(t) = \mathrm{e}^{-i\omega_1 t \mathbf{I_x}} \rho(0) \mathrm{e}^{+i\omega_1 t \mathbf{I_x}} \tag{8.26}$$

with

$$\rho(0) = \mathbf{I_z}\,. \tag{8.27}$$

To make further progress we have to find expressions for the exponential matrices in Eqn 8.26. There are several possible approaches, but the most generally practical method, described in Appendix 8.2, uses the properties of diagonal matrices. Thus,

Details of this calculation are given in Appendix 8.3.

$$\mathrm{e}^{\pm i\omega_1 t \mathbf{I_x}} = \begin{pmatrix} \cos(\tfrac{1}{2}\omega_1 t) & \pm i \sin(\tfrac{1}{2}\omega_1 t) \\ \pm i \sin(\tfrac{1}{2}\omega_1 t) & \cos(\tfrac{1}{2}\omega_1 t) \end{pmatrix}. \tag{8.28}$$

We can then evaluate $\rho(t)$ by multiplying matrices according to Eqn 8.26 to obtain

$$\rho(t) = \begin{pmatrix} \tfrac{1}{2}\cos\omega_1 t & \tfrac{1}{2} i \sin\omega_1 t \\ -\tfrac{1}{2} i \sin\omega_1 t & -\tfrac{1}{2}\cos\omega_1 t \end{pmatrix}. \tag{8.29}$$

Finally, we can use Eqn 8.4 to calculate expectation values:

$$\begin{aligned}
\langle \hat{I}_x \rangle &= \mathrm{Tr}(\rho \mathbf{I_x}) = \mathrm{Tr}\begin{pmatrix} \tfrac{1}{4} i \sin\omega_1 t & \tfrac{1}{4}\cos\omega_1 t \\ -\tfrac{1}{4}\cos\omega_1 t & -\tfrac{1}{4} i \sin\omega_1 t \end{pmatrix} = 0 \\
\langle \hat{I}_y \rangle &= \mathrm{Tr}(\rho \mathbf{I_y}) = \mathrm{Tr}\begin{pmatrix} -\tfrac{1}{4}\sin\omega_1 t & -\tfrac{1}{4} i \cos\omega_1 t \\ -\tfrac{1}{4} i \cos\omega_1 t & -\tfrac{1}{4}\sin\omega_1 t \end{pmatrix} = -\tfrac{1}{2}\sin\omega_1 t \\
\langle \hat{I}_z \rangle &= \mathrm{Tr}(\rho \mathbf{I_z}) = \mathrm{Tr}\begin{pmatrix} \tfrac{1}{4}\cos\omega_1 t & -\tfrac{1}{4} i \sin\omega_1 t \\ -\tfrac{1}{4} i \sin\omega_1 t & \tfrac{1}{4}\cos\omega_1 t \end{pmatrix} = \tfrac{1}{2}\cos\omega_1 t.
\end{aligned} \tag{8.30}$$

As expected from the vector model, a radiofrequency pulse along the x axis causes the expectation values of $\hat{I}_y$ and $\hat{I}_z$ to oscillate, at frequency ω_1, while the expectation value of $\hat{I}_x$ remains constant at zero.

cf. Fig. 1.2

8.6 Connection to product operators

The approach just outlined is ideal for numerical calculations on a computer, especially for more complex problems where analytical expressions for matrix exponentials are hard or impossible to obtain. In some cases, however, there is a more subtle and convenient method which exploits the properties of the operators involved rather than relying on somewhat tiresome matrix arithmetic. It also gives insight into the link between product operators and density matrices.

Consider a system described by an initial density operator at $t = 0$,

From now on we use operators (e.g. $\hat{A}$) and their matrix representations ($\mathbf{A}$) interchangeably.

$$\rho(0) = \mathbf{A} \tag{8.31}$$

(e.g. $\mathbf{A} = \mathbf{I_z}$, as in the previous section) and imagine that we wish to determine the evolution of a spin system under the influence of a Hamiltonian

$$\mathbf{H} = b\mathbf{B} \tag{8.32}$$

(e.g. $b = \omega_1$, $\mathbf{B} = \mathbf{I_x}$, as in Eqn 8.25). Suppose that the following commutation relations hold:

$$\begin{aligned} [\mathbf{A},\mathbf{B}] &= i\mathbf{C}, \\ [\mathbf{B},\mathbf{C}] &= i\mathbf{A}. \end{aligned} \tag{8.33}$$

Under these admittedly special circumstances, the evolution of the density operator is particularly simple:

$$\rho(t) = \mathbf{A}\cos bt - \mathbf{C}\sin bt\ . \tag{8.34}$$

This important relation can be justified as follows. With the definition of $\mathbf{H}$ given above, the Liouville–von Neumann equation is

$$\frac{\mathrm{d}\rho}{\mathrm{d}t} = -ib[\mathbf{B},\rho]\ . \tag{8.35}$$

Substituting Eqn 8.34 into the right-hand side gives:

$$\begin{aligned} -ib[\mathbf{B},\rho] &= -ib[\mathbf{B},(\mathbf{A}\cos bt - \mathbf{C}\sin bt)] \\ &= -ib[\mathbf{B},\mathbf{A}]\cos bt + ib[\mathbf{B},\mathbf{C}]\sin bt \\ &= -ib(-i\mathbf{C})\cos bt + ib(i\mathbf{A})\sin bt \\ &= -b\mathbf{C}\cos bt - b\mathbf{A}\sin bt, \end{aligned} \tag{8.36}$$

which is precisely the same as the left-hand side of Eqn 8.35:

$$\frac{\mathrm{d}\rho}{\mathrm{d}t} = \frac{\mathrm{d}}{\mathrm{d}t}\left(\mathbf{A}\cos bt - \mathbf{C}\sin bt\right) = -b\mathbf{A}\sin bt - b\mathbf{C}\cos bt. \tag{8.37}$$

As a check that Eqn 8.34 is applicable, the relevant commutators are:

$[\mathbf{A},\mathbf{B}] = [\mathbf{I_z},\mathbf{I_x}] = i\mathbf{I_y} = i\mathbf{C}$;

$[\mathbf{B},\mathbf{C}] = [\mathbf{I_x},\mathbf{I_y}] = i\mathbf{I_z} = i\mathbf{A}$.

To see how Eqn 8.34 works, we first repeat the calculation of Section 8.5 (evolution during an on-resonance radiofrequency pulse). The three operators involved are $\mathbf{A} = \mathbf{I_z}$, $\mathbf{B} = \mathbf{I_x}$, $\mathbf{C} = \mathbf{I_y}$, with $b = \omega_1$. Thus,

$$\rho(t) = \mathbf{I_z}\cos\omega_1 t - \mathbf{I_y}\sin\omega_1 t$$

$$= \begin{pmatrix} \frac{1}{2} & 0 \\ 0 & -\frac{1}{2} \end{pmatrix}\cos\omega_1 t - \begin{pmatrix} 0 & -\frac{1}{2}i \\ \frac{1}{2}i & 0 \end{pmatrix}\sin\omega_1 t = \begin{pmatrix} \frac{1}{2}\cos\omega_1 t & \frac{1}{2}i\sin\omega_1 t \\ -\frac{1}{2}i\sin\omega_1 t & -\frac{1}{2}\cos\omega_1 t \end{pmatrix} \quad (8.38)$$

as in Eqn 8.29.

Now consider the free precession that occurs after a $90°_x$ pulse: $\mathbf{A} = -\mathbf{I_y}$, $\mathbf{B} = \mathbf{I_z}$, $\mathbf{C} = -\mathbf{I_x}$, and $b = \Omega$ so that

$$\rho(t) = -\mathbf{I_y}\cos\Omega t + \mathbf{I_x}\sin\Omega t\,, \quad (8.39)$$

again as expected (Figs 1.4 and 3.2).

Expectation values can be obtained directly from equations such as 8.39, e.g.

$$\langle \hat{I}_y \rangle = \mathrm{Tr}[\rho\mathbf{I_y}] = -\mathrm{Tr}[\mathbf{I_y^2}]\cos\Omega t + \mathrm{Tr}[\mathbf{I_x}\mathbf{I_y}]\sin\Omega t$$

$$= -\mathrm{Tr}\begin{pmatrix} \frac{1}{4} & 0 \\ 0 & \frac{1}{4} \end{pmatrix}\cos\Omega t + \mathrm{Tr}\begin{pmatrix} \frac{1}{4}i & 0 \\ 0 & -\frac{1}{4}i \end{pmatrix}\sin\Omega t = -\frac{1}{2}\cos\Omega t, \quad (8.40)$$

and similarly $\langle \hat{I}_x \rangle = \sin(\Omega t)/2$.

All of this should be looking pretty familiar from Chapters 3–5 of Part A. If we replace the matrices by the corresponding product operators, Eqn 8.39 is *exactly* the result of the product operator formalism. To emphasize the connection between the two approaches, we can rewrite Eqn 8.34 as:

$$\mathbf{A} \xrightarrow{b\mathbf{B}t} \mathbf{A}\cos bt - \mathbf{C}\sin bt\,. \quad (8.41)$$

Thus, product operators are nothing more than a cunning way of doing analytical density matrix calculations.

The commutators are:
$[\mathbf{A},\mathbf{B}] = [-\mathbf{I_y},\mathbf{I_z}] = -i\mathbf{I_x} = i\mathbf{C}$;
$[\mathbf{B},\mathbf{C}] = [\mathbf{I_z}, -\mathbf{I_x}] = -i\mathbf{I_y} = i\mathbf{A}$.

Note that $\mathrm{Tr}[\mathbf{I_j}\mathbf{I_k}] = 1/2$ if $j = k$, or 0 if $j \neq k$ ($j, k = x, y, z$)

The correspondence of product operators and density matrices, and the use of Eqns 8.34 and 8.41 will be pursued further in Chapter 9. By convention, product operators do not have 'hats', i.e. I_x not $\hat{I}_x$.

Appendix 8.1. Pure and mixed states, and coherence

To see the difference between pure and mixed states consider two different ensembles. In both, the microscopic systems are described by

$$|\psi\rangle = a_\alpha e^{i\phi_\alpha}|\alpha\rangle + a_\beta e^{i\phi_\beta}|\beta\rangle\,, \quad (8.42)$$

where a_α and a_β are amplitudes, and ϕ_α and ϕ_β are phases, and all four quantities are real. The corresponding density matrix is

$$\rho = \begin{pmatrix} a_\alpha^2 & a_\alpha a_\beta e^{i(\phi_\alpha - \phi_\beta)} \\ a_\alpha a_\beta e^{-i(\phi_\alpha - \phi_\beta)} & a_\beta^2 \end{pmatrix}. \quad (8.43)$$

In the first ensemble (a *pure* state) all the systems are in the same state, and so the density matrix of the whole ensemble is identical to that of each of the microscopic systems. The second ensemble (a *mixed* state) is similar to the first, except that the microscopic systems are in subtly different states, with the phase parameters, ϕ_α and ϕ_β, varying randomly from one system to the next. The macroscopic density matrix is obtained by averaging over the ensemble, i.e. over all values of ϕ_α and ϕ_β. The (random) phase factors $\exp[\pm i(\phi_\alpha - \phi_\beta)]$ interfere destructively to give

$$\rho = \begin{pmatrix} a_\alpha^2 & 0 \\ 0 & a_\beta^2 \end{pmatrix}. \tag{8.44}$$

Comparing these two examples, we see that the diagonal elements of the density matrices, which correspond to the populations of the two basis states, are the same in both cases, but the *off-diagonal* elements are quite different. These off-diagonal elements, which arise from systems in superposition states, are only non-zero when there is *phase coherence* between the states of the microscopic systems. Such *phase coherent superpositions* are called *coherences*. In NMR, they are created by applying one or more pulses of coherent radiofrequency radiation.

Appendix 8.2. Matrix diagonalization and matrix exponentials

When a matrix is diagonal (i.e. all the elements off the main diagonal are zero) it is particularly simple to evaluate matrix powers. Consider a general 2×2 diagonal matrix

$$\mathbf{D} = \begin{pmatrix} a & 0 \\ 0 & b \end{pmatrix} \tag{8.45}$$

for which

$$\mathbf{D}^2 = \begin{pmatrix} a & 0 \\ 0 & b \end{pmatrix}\begin{pmatrix} a & 0 \\ 0 & b \end{pmatrix} = \begin{pmatrix} a^2 & 0 \\ 0 & b^2 \end{pmatrix}. \tag{8.46}$$

Similar expressions can be written for any power of **D**, and thus for the matrix exponential (which is a sum of powers of **D**):

$$e^{\mathbf{D}} = 1 + \mathbf{D} + \frac{\mathbf{D}^2}{2!} + \frac{\mathbf{D}^3}{3!} + \cdots = \begin{pmatrix} e^a & 0 \\ 0 & e^b \end{pmatrix}. \tag{8.47}$$

Most matrices, such as $\mathbf{I}_x$, are not diagonal, and this simple approach cannot be applied. However, it is possible to do something very similar by *diagonalizing* the matrix.

A matrix **M** can be diagonalized by finding matrices **S** and **Λ** such that

$$\mathbf{M} = \mathbf{S\Lambda S}^{-1}, \tag{8.48}$$

where **Λ** is diagonal. Then

Using $\mathbf{S}^{-1}\mathbf{S} = \mathbf{1}$.

$$\mathbf{M}^2 = \mathbf{S\Lambda S}^{-1}\mathbf{S\Lambda S}^{-1} = \mathbf{S\Lambda\Lambda S}^{-1} = \mathbf{S\Lambda}^2\mathbf{S}^{-1} \tag{8.49}$$

and so on. Thus,

$$e^{\mathbf{M}} = \mathbf{S}e^{\mathbf{\Lambda}}\mathbf{S}^{-1}. \tag{8.50}$$

A matrix can be diagonalized by finding its eigenvectors and eigenvalues. Suppose we have an $n \times n$ square matrix $\mathbf{M}$ and a vector $\mathbf{x}$ such that

$$\mathbf{Mx} = \lambda \mathbf{x} , \tag{8.51}$$

where λ is a number; $\mathbf{x}$ is said to be an eigenvector of $\mathbf{M}$ with eigenvalue λ. In general there will be n eigenvectors, each with its own eigenvalue. One way to determine them is to rewrite the eigenvalue equation as

$$(\mathbf{M} - \lambda \mathbf{1})\mathbf{x} = \mathbf{0} , \tag{8.52}$$

where $\mathbf{0}$ is a vector consisting entirely of zeroes and $\mathbf{1}$ is the unit matrix. This set of n simultaneous equations only has non-trivial solutions if the determinant of the coefficients disappears, that is if

The trivial solution is $\mathbf{x} = \mathbf{0}$.

$$|\mathbf{M} - \lambda \mathbf{1}| = 0 . \tag{8.53}$$

This expression is called the *secular equation*; expansion of the determinant gives a polynomial with n roots, corresponding to the n eigenvalues. This method works well for analytical calculations with small matrices, but for large matrices it is usually more efficient to use numerical methods. These can be found in many texts on numerical analysis (e.g. Press *et al*. 1992).

Once the set of eigenvalues and eigenvectors has been found, the matrices $\mathbf{S}$ and $\boldsymbol{\Lambda}$ can readily be determined: $\boldsymbol{\Lambda}$ contains the eigenvalues of $\mathbf{M}$ on its diagonal (and zeroes everywhere else) and the columns of $\mathbf{S}$ are the corresponding eigenvectors.

Finally we prove a few useful properties of commuting matrices, i.e. matrices for which $[\mathbf{A},\mathbf{B}] = \mathbf{AB} - \mathbf{BA} = \mathbf{0}$. First:

$$\left[e^{\mathbf{A}},\mathbf{B}\right] = \left[e^{\mathbf{B}},\mathbf{A}\right] = \left[e^{\mathbf{A}},e^{\mathbf{B}}\right] = \mathbf{0} \tag{8.54}$$

which follows from the fact that $e^{\mathbf{A}}$ is a sum of powers of $\mathbf{A}$, and $\mathbf{B}$ commutes with $\mathbf{A}$ and so with any power of $\mathbf{A}$.

Second:

$$e^{\mathbf{A}+\mathbf{B}} = e^{\mathbf{A}}e^{\mathbf{B}} = e^{\mathbf{B}}e^{\mathbf{A}} . \tag{8.55}$$

This can be shown by expanding $e^{\mathbf{A}}$ and $e^{\mathbf{B}}$

$$\begin{aligned} e^{\mathbf{A}}e^{\mathbf{B}} &= \left(\mathbf{1}+\mathbf{A}+\frac{\mathbf{A}^2}{2!}+\frac{\mathbf{A}^3}{3!}+\cdots\right)\left(\mathbf{1}+\mathbf{B}+\frac{\mathbf{B}^2}{2!}+\frac{\mathbf{B}^3}{3!}+\cdots\right) \\ &= \mathbf{1}+\mathbf{A}+\mathbf{B}+\frac{\mathbf{A}^2}{2!}+\mathbf{AB}+\frac{\mathbf{B}^2}{2!}+\frac{\mathbf{A}^3}{3!}+\frac{\mathbf{A}^2\mathbf{B}}{2!}+\frac{\mathbf{AB}^2}{2!}+\frac{\mathbf{B}^3}{3!}+\cdots \\ &= \mathbf{1}+(\mathbf{A}+\mathbf{B})+\frac{(\mathbf{A}+\mathbf{B})^2}{2!}+\frac{(\mathbf{A}+\mathbf{B})^3}{3!}+\cdots \\ &= e^{\mathbf{A}+\mathbf{B}}, \end{aligned} \tag{8.56}$$

where the commutator $[\mathbf{A},\mathbf{B}] = \mathbf{0}$ has been used to write

$$\mathbf{A}^2 + 2\mathbf{AB} + \mathbf{B}^2 = \mathbf{A}^2 + \mathbf{AB} + \mathbf{BA} + \mathbf{B}^2 \equiv (\mathbf{A}+\mathbf{B})^2 , \tag{8.57}$$

and so on.

This property was used in Eqn 8.15.

Third, note that for any matrix $[e^{\mathbf{A}}, \mathbf{A}] = \mathbf{0}$.

Equations 8.54 and 8.55 lead to a considerable and very welcome simplification in some density matrix calculations. If the Hamiltonian is the sum of two commuting parts ($\mathbf{H} = \mathbf{H_1} + \mathbf{H_2}$), then the evolution predicted by the Liouville–von Neumann equation may be written

$$\begin{aligned} \rho(t) &= e^{-i\mathbf{H}t}\rho(0)e^{i\mathbf{H}t} \\ &= e^{-i\mathbf{H}_1 t}e^{-i\mathbf{H}_2 t}\rho(0)e^{i\mathbf{H}_2 t}e^{i\mathbf{H}_1 t} \\ &= e^{-i\mathbf{H}_2 t}e^{-i\mathbf{H}_1 t}\rho(0)e^{i\mathbf{H}_1 t}e^{i\mathbf{H}_2 t}, \end{aligned} \tag{8.58}$$

i.e. the evolution produced by $\mathbf{H_1}$ and $\mathbf{H_2}$ may be calculated sequentially, rather than simultaneously, and in any order (a property that holds regardless of how many commuting parts constitute the overall Hamiltonian). This result underlies the whole of the product operator formalism for weakly coupled spin systems.

Appendix 8.3. The matrix exponential of $\hat{I}_x$

The matrix exponential of $\hat{I}_x$ can be determined using the approach described in Appendix 8.2. Let $\mathbf{M}$ be $k\mathbf{I_x}$, where k is some (real or complex) number. The secular equation is

$$\begin{vmatrix} -\lambda & \frac{1}{2}k \\ \frac{1}{2}k & -\lambda \end{vmatrix} = \lambda^2 - \tfrac{1}{4}k^2 = 0\,. \tag{8.59}$$

Thus, the two eigenvalues are $\lambda = \pm k/2$.

The eigenvectors may be deduced by substituting each of the eigenvalues in turn into the eigenvalue equation, $(\mathbf{M} - \lambda\mathbf{1})\mathbf{x} = \mathbf{0}$. For $\lambda = k/2$

$$\begin{pmatrix} -\frac{1}{2}k & \frac{1}{2}k \\ \frac{1}{2}k & -\frac{1}{2}k \end{pmatrix}\begin{pmatrix} x_1 \\ x_2 \end{pmatrix} = \begin{pmatrix} 0 \\ 0 \end{pmatrix} \tag{8.60}$$

which has the solution $x_1 = x_2$. Note that the eigenvector is not completely defined: we could multiply x_1 and x_2 by any number and still have an eigenvector. One common convention is to normalize the eigenvectors to unit length; for the case considered above this gives $x_1 = x_2 = 1/\sqrt{2}$. The second eigenvector can be determined in the same way by substituting $\lambda = -k/2$, giving $x_1 = 1/\sqrt{2}$, $x_2 = -1/\sqrt{2}$.

The matrices $\mathbf{\Lambda}$ and $\mathbf{S}$ are thus (Appendix 8.2)

$$\mathbf{\Lambda} = \begin{pmatrix} \frac{1}{2}k & 0 \\ 0 & -\frac{1}{2}k \end{pmatrix}, \qquad \mathbf{S} = \frac{1}{\sqrt{2}}\begin{pmatrix} 1 & 1 \\ 1 & -1 \end{pmatrix}. \tag{8.61}$$

Now we can calculate the matrix exponential of $\mathbf{M}$, with $k = \pm i\omega_1 t$ (as in Section 8.5):

$$e^{\pm i\omega_1 t\mathbf{I}_x} = \mathbf{S}e^{\Lambda}\mathbf{S}^{-1} = \frac{1}{2}\begin{pmatrix}1 & 1\\ 1 & -1\end{pmatrix}\begin{pmatrix}e^{\pm i\omega_1 t/2} & 0\\ 0 & e^{\mp i\omega_1 t/2}\end{pmatrix}\begin{pmatrix}1 & 1\\ 1 & -1\end{pmatrix}$$
$$= \begin{pmatrix}\cos(\frac{1}{2}\omega_1 t) & \pm i\sin(\frac{1}{2}\omega_1 t)\\ \pm i\sin(\frac{1}{2}\omega_1 t) & \cos(\frac{1}{2}\omega_1 t)\end{pmatrix} \quad (8.62)$$

N.B. $\mathbf{S}^{-1} = \mathbf{S}$ here.

as stated in Eqn 8.28.

A more elegant proof is possible here because $\mathbf{I_x}$ has the special property that its square is proportional to the unit matrix, **1** (Eqn 7.28). If a matrix **A** is such that $\mathbf{A}^2 = \mathbf{1}$ then

$$e^{ia\mathbf{A}} \equiv 1 + ia\mathbf{A} + \frac{i^2a^2\mathbf{A}^2}{2!} + \frac{i^3a^3\mathbf{A}^3}{3!} + \cdots \quad (8.63)$$

may be rewritten, using $\mathbf{A} = \mathbf{A}^3 = \mathbf{A}^5 = \cdots$ and $\mathbf{1} = \mathbf{A}^2 = \mathbf{A}^4 = \cdots$, as

$$\mathbf{1}\left(1 - \frac{a^2}{2!} + \frac{a^4}{4!} - \cdots\right) + i\mathbf{A}\left(a - \frac{a^3}{3!} + \frac{a^5}{5!} - \cdots\right). \quad (8.64)$$

The two series in parentheses are simply the expansions of cos a and sin a, so that

$$e^{ia\mathbf{A}} = \mathbf{1}\cos a + i\mathbf{A}\sin a\,. \quad (8.65)$$

To evaluate $\exp[\pm i\omega_1 t\mathbf{I_x}]$, we replace **A** by $2\mathbf{I_x}$ (because $4\mathbf{I_x}^2 = \mathbf{1}$) and a by $\pm\omega_1 t/2$ to obtain

$$e^{\pm i\omega_1 t\mathbf{I}_x} = \mathbf{1}\cos\left(\tfrac{1}{2}\omega_1 t\right) \pm i2\mathbf{I_x}\sin\left(\tfrac{1}{2}\omega_1 t\right) = \begin{pmatrix}\cos(\frac{1}{2}\omega_1 t) & \pm i\sin(\frac{1}{2}\omega_1 t)\\ \pm i\sin(\frac{1}{2}\omega_1 t) & \cos(\frac{1}{2}\omega_1 t)\end{pmatrix}. \quad (8.66)$$

Appendix 8.4. The rotating frame

The rotating-frame transformation plays a central role in the theory of many NMR experiments, greatly simplifying calculations by removing the continual rotation at the Larmor frequency, which clutters things in the laboratory frame.

Radiofrequency irradiation of an NMR sample creates a linearly oscillating magnetic field, which can be decomposed into the sum of two counter-rotating magnetic fields:

$$\cos\omega_{\mathrm{rf}}t = \tfrac{1}{2}\left(e^{+i\omega_{\mathrm{rf}}t} + e^{-i\omega_{\mathrm{rf}}t}\right). \quad (8.67)$$

If the oscillating field is nearly resonant with the NMR transition, one component rotates at a frequency ω_{rf}, close to the Larmor frequency, while the other is at $-\omega_{\mathrm{rf}}$. The latter will have little effect and can be ignored. Thus the effective Hamiltonian in the laboratory frame can be written as

$$\hat{H} = \omega_0\hat{I}_z + \omega_1\left(\hat{I}_x\cos\omega_{\mathrm{rf}}t + \hat{I}_y\sin\omega_{\mathrm{rf}}t\right), \quad (8.68)$$

where the phase of the radiofrequency radiation has been chosen such that it initially lies along the x axis of the laboratory frame. Because this Hamiltonian varies rapidly with time it is difficult to calculate its effect on a spin system. However, as the time variation has a simple regular form it is possible to remove it by transferring to a coordinate system in synchrony with the radiofrequency field, as outlined in Section 1.3. This may be seen as follows.

The evolution of a wavefunction under a Hamiltonian is given by the time-dependent Schrödinger equation,

$$\frac{\mathrm{d}}{\mathrm{d}t}|\psi\rangle = -i\hat{H}|\psi\rangle . \tag{8.69}$$

Let us define a modified wavefunction, $|\psi_\mathrm{R}\rangle$, related to $|\psi\rangle$ by rotation around the z axis at a rate ω_rf,

$$|\psi_\mathrm{R}\rangle = \mathrm{e}^{i\omega_\mathrm{rf}t\hat{I}_z}|\psi\rangle . \tag{8.70}$$

The time-dependence of $|\psi_\mathrm{R}\rangle$ can be written in terms of a modified Hamiltonian as follows:

$$\begin{aligned}\frac{\mathrm{d}}{\mathrm{d}t}|\psi_\mathrm{R}\rangle &= i\omega_\mathrm{rf}\hat{I}_z\mathrm{e}^{i\omega_\mathrm{rf}t\hat{I}_z}|\psi\rangle + \mathrm{e}^{i\omega_\mathrm{rf}t\hat{I}_z}\frac{\mathrm{d}}{\mathrm{d}t}|\psi\rangle \\ &= i\omega_\mathrm{rf}\hat{I}_z|\psi_\mathrm{R}\rangle + \mathrm{e}^{i\omega_\mathrm{rf}t\hat{I}_z}(-i\hat{H})|\psi\rangle \\ &= -i\left[-\omega_\mathrm{rf}\hat{I}_z|\psi_\mathrm{R}\rangle + \mathrm{e}^{i\omega_\mathrm{rf}t\hat{I}_z}\hat{H}\mathrm{e}^{-i\omega_\mathrm{rf}t\hat{I}_z}|\psi_\mathrm{R}\rangle\right] \\ &= -i\left[-\omega_\mathrm{rf}\hat{I}_z + \mathrm{e}^{i\omega_\mathrm{rf}t\hat{I}_z}\hat{H}\mathrm{e}^{-i\omega_\mathrm{rf}t\hat{I}_z}\right]|\psi_\mathrm{R}\rangle \\ &\equiv -i\hat{H}_\mathrm{R}|\psi_\mathrm{R}\rangle .\end{aligned} \tag{8.71}$$

The third line follows from the second using $|\psi\rangle = \exp[-i\omega_\mathrm{rf}t\hat{I}_z]|\psi_\mathrm{R}\rangle$.

$\hat{H}_\mathrm{R}$ is made up of two parts: the original Hamiltonian subjected to the same rotation as the wavefunction, and an additional term, $-\omega_\mathrm{rf}\hat{I}_z$, lying along the rotation axis. For the laboratory frame Hamiltonian above,

$$\begin{aligned}\hat{H}_\mathrm{R} &= -\omega_\mathrm{rf}\hat{I}_z + \mathrm{e}^{i\omega_\mathrm{rf}t\hat{I}_z}\hat{H}\mathrm{e}^{-i\omega_\mathrm{rf}t\hat{I}_z} \\ &= -\omega_\mathrm{rf}\hat{I}_z + \omega_0\hat{I}_z + \omega_1\hat{I}_x \\ &= (\omega_0 - \omega_\mathrm{rf})\hat{I}_z + \omega_1\hat{I}_x .\end{aligned} \tag{8.72}$$

The second line can be deduced using Eqn 8.34.

Thus, the time dependence of the radiofrequency field has been removed and the precession frequency around the $\mathbf{B_0}$ direction has been reduced from ω_0 to $\omega_0 - \omega_\mathrm{rf} = \Omega$. If the radiofrequency field ω_1 is much larger than the resonance offset Ω, the rotating frame Hamiltonian becomes particularly simple:

$$\hat{H}_R = \omega_1\hat{I}_x . \tag{8.73}$$

The same result is obtained if the radiofrequency field is in exact resonance with the Larmor frequency, $\omega_0 = \omega_\mathrm{rf}$.

9 Weak coupling and equivalence

9.1 Introduction

We have now seen all the machinery needed to use density matrices to predict the behaviour of isolated nuclei. As in the case of product operators, the density matrix approach reassuringly reproduces the results of the vector model. With two or more coupled spins, however, density matrices start to come into their own. Here we discuss two situations—weakly coupled and equivalent spins—to see more clearly the correspondence between product operators and density matrices and to derive some of the results quoted without proof in Part A.

9.2 Density operators in two-spin systems

The methods outlined in the previous two chapters can easily be extended to a two-spin (IS) system, which will have four basis vectors:

$$|\psi\rangle = c_{\alpha\alpha}|\alpha_{\mathrm{I}}\alpha_{\mathrm{S}}\rangle + c_{\alpha\beta}|\alpha_{\mathrm{I}}\beta_{\mathrm{S}}\rangle + c_{\beta\alpha}|\beta_{\mathrm{I}}\alpha_{\mathrm{S}}\rangle + c_{\beta\beta}|\beta_{\mathrm{I}}\beta_{\mathrm{S}}\rangle . \tag{9.1}$$

Matrix representations of the six one-spin angular momentum operators, $\mathbf{I_x}$, $\mathbf{I_y}$, $\mathbf{I_z}$, $\mathbf{S_x}$, $\mathbf{S_y}$ and $\mathbf{S_z}$, can be obtained using Eqns 7.21 and 7.25, e.g.

$$\hat{I}_x|\alpha_{\mathrm{I}}\alpha_{\mathrm{S}}\rangle = \tfrac{1}{2}|\beta_{\mathrm{I}}\alpha_{\mathrm{S}}\rangle, \qquad \hat{S}_y|\beta_{\mathrm{I}}\alpha_{\mathrm{S}}\rangle = \tfrac{1}{2}i|\beta_{\mathrm{I}}\beta_{\mathrm{S}}\rangle \tag{9.2}$$

(note that I operators do not affect spin S and vice versa). Thus, for example,

$$\mathbf{I_x} = \begin{pmatrix} 0 & 0 & \frac{1}{2} & 0 \\ 0 & 0 & 0 & \frac{1}{2} \\ \frac{1}{2} & 0 & 0 & 0 \\ 0 & \frac{1}{2} & 0 & 0 \end{pmatrix}. \tag{9.3}$$

There is, however, a simple approach which allows matrix representations to be deduced directly. Two-spin matrices can be obtained by taking *direct products* of one-spin matrices (Eqns 7.23 and 7.26) and the unit matrix. Thus,

Direct products are explained in Appendix 9.1.

$$\mathbf{I_y} = \begin{pmatrix} 0 & -\frac{1}{2}i \\ \frac{1}{2}i & 0 \end{pmatrix} \otimes \begin{pmatrix} 1 & 0 \\ 0 & 1 \end{pmatrix} = \begin{pmatrix} 0 & 0 & -\frac{1}{2}i & 0 \\ 0 & 0 & 0 & -\frac{1}{2}i \\ \frac{1}{2}i & 0 & 0 & 0 \\ 0 & \frac{1}{2}i & 0 & 0 \end{pmatrix} \tag{9.4}$$

and so on. Matrices corresponding to spin S are calculated in the same way, except that the order of the two matrices in the direct product is reversed:

$$\mathbf{S_y} = \begin{pmatrix} 1 & 0 \\ 0 & 1 \end{pmatrix} \otimes \begin{pmatrix} 0 & -\frac{1}{2}i \\ \frac{1}{2}i & 0 \end{pmatrix} = \begin{pmatrix} 0 & -\frac{1}{2}i & 0 & 0 \\ \frac{1}{2}i & 0 & 0 & 0 \\ 0 & 0 & 0 & -\frac{1}{2}i \\ 0 & 0 & \frac{1}{2}i & 0 \end{pmatrix}. \tag{9.5}$$

Note that, as expected, all I matrices commute with all S matrices, e.g. $[\mathbf{I_x}, \mathbf{S_y}] = \mathbf{0}$. Moreover, the commutation and other relations given in Eqns 7.28 – 7.30 still hold, e.g. $\mathbf{I_x}^2 = \mathbf{1/4}$, $\mathbf{I_x I_y} = i\mathbf{I_z}/2$, $[\mathbf{I_x}, \mathbf{I_y}] = i\mathbf{I_z}$.

Matrix representations of the other two-spin operators can be obtained in much the same way, either by taking direct products of the corresponding one-spin matrices,

$$2\mathbf{I_x S_y} = 2 \times \begin{pmatrix} 0 & \frac{1}{2} \\ \frac{1}{2} & 0 \end{pmatrix} \otimes \begin{pmatrix} 0 & -\frac{1}{2}i \\ \frac{1}{2}i & 0 \end{pmatrix} = \begin{pmatrix} 0 & 0 & 0 & -\frac{1}{2}i \\ 0 & 0 & \frac{1}{2}i & 0 \\ 0 & -\frac{1}{2}i & 0 & 0 \\ \frac{1}{2}i & 0 & 0 & 0 \end{pmatrix}, \tag{9.6}$$

or by multiplying the corresponding two-spin matrices,

$$2\mathbf{I_y S_x} = 2 \times \begin{pmatrix} 0 & 0 & -\frac{1}{2}i & 0 \\ 0 & 0 & 0 & -\frac{1}{2}i \\ \frac{1}{2}i & 0 & 0 & 0 \\ 0 & \frac{1}{2}i & 0 & 0 \end{pmatrix} \begin{pmatrix} 0 & \frac{1}{2} & 0 & 0 \\ \frac{1}{2} & 0 & 0 & 0 \\ 0 & 0 & 0 & \frac{1}{2} \\ 0 & 0 & \frac{1}{2} & 0 \end{pmatrix} = \begin{pmatrix} 0 & 0 & 0 & -\frac{1}{2}i \\ 0 & 0 & -\frac{1}{2}i & 0 \\ 0 & \frac{1}{2}i & 0 & 0 \\ \frac{1}{2}i & 0 & 0 & 0 \end{pmatrix}. \tag{9.7}$$

For convenience we give a complete list of the two-spin matrices in Appendix 9.2.

It was noted in Chapter 8 that the off-diagonal elements of the density matrix represent the *coherences* we have previously categorized as zero, single, double, etc., according to the difference in magnetic quantum number, m, of the two states involved. This may be seen more clearly from the matrix representations of the two-spin operators. For example, the off-diagonal elements of $\mathbf{I_x}$ (Eqn 9.3) connect pairs of states such as $|\alpha_I\alpha_S\rangle$ and $|\beta_I\alpha_S\rangle$, for which $\Delta m = \pm 1$, i.e. single-quantum coherences. Operator products such as $2\mathbf{I_x S_y}$ in Eqn 9.6 have matrix elements in the off-diagonal corners which connect $|\alpha_I\alpha_S\rangle$ and $|\beta_I\beta_S\rangle$ ($\Delta m = \pm 2$, double-quantum coherence) as well as elements linking $|\alpha_I\beta_S\rangle$ and $|\beta_I\alpha_S\rangle$ ($\Delta m = 0$, zero-quantum coherence).

Eqn 4.5 gives the product operator forms of zero- and double-quantum coherence.

9.3 *J* coupling

The Hamiltonian describing the J (scalar) coupling between spins I and S has the general form

$$2\pi J \hat{I} \cdot \hat{S} \tag{9.8}$$

Henceforth the symbol J_{IS} is abbreviated to J.

where J, the spin–spin coupling constant, is the strength of the coupling (coupling constants are traditionally measured in Hz; it is therefore necessary to multiply by 2π to convert them into angular frequency units). The scalar product can be expanded as the sum of three terms

$$\hat{I}\cdot\hat{S}=\hat{I}_x\hat{S}_x+\hat{I}_y\hat{S}_y+\hat{I}_z\hat{S}_z \tag{9.9}$$

so that the matrix form of the coupling Hamiltonian is

$$2\pi J\left(\mathbf{I_xS_x}+\mathbf{I_yS_y}+\mathbf{I_zS_z}\right)=2\pi J\begin{pmatrix}\frac{1}{4}&0&0&0\\0&-\frac{1}{4}&\frac{1}{2}&0\\0&\frac{1}{2}&-\frac{1}{4}&0\\0&0&0&\frac{1}{4}\end{pmatrix}. \tag{9.10}$$

Evolution under the scalar coupling does not occur in isolation; one must consider the entire Hamiltonian, including the (Zeeman) interactions with the magnetic field. For a system of two coupled spins the complete Hamiltonian in the rotating frame has the form

$$\hat{H}_{\mathrm{IS}}=\Omega_{\mathrm{I}}\hat{I}_z+\Omega_{\mathrm{S}}\hat{S}_z+2\pi J\,\hat{I}\cdot\hat{S} \tag{9.11}$$

or

$$\mathbf{H_{IS}}=\begin{pmatrix}\frac{1}{2}(\Omega_{\mathrm{I}}+\Omega_{\mathrm{S}})+\frac{1}{2}\pi J&0&0&0\\0&\frac{1}{2}(\Omega_{\mathrm{I}}-\Omega_{\mathrm{S}})-\frac{1}{2}\pi J&\pi J&0\\0&\pi J&-\frac{1}{2}(\Omega_{\mathrm{I}}-\Omega_{\mathrm{S}})-\frac{1}{2}\pi J&0\\0&0&0&-\frac{1}{2}(\Omega_{\mathrm{I}}+\Omega_{\mathrm{S}})+\frac{1}{2}\pi J\end{pmatrix} \tag{9.12}$$

In order to determine the evolution produced by $\mathbf{H_{IS}}$ it is necessary to calculate its matrix exponential (see Chapter 10). Before doing this, however, it is useful to consider an approximate approach.

9.4 Weak coupling: a brute force approach

The Hamiltonian matrix described by Eqn 9.12 is almost diagonal. The weak coupling approximation simply neglects these two elements and uses the simpler form

$$\mathbf{H'_{IS}}=\begin{pmatrix}\frac{1}{2}(\Omega_{\mathrm{I}}+\Omega_{\mathrm{S}})+\frac{1}{2}\pi J&0&0&0\\0&\frac{1}{2}(\Omega_{\mathrm{I}}-\Omega_{\mathrm{S}})-\frac{1}{2}\pi J&0&0\\0&0&-\frac{1}{2}(\Omega_{\mathrm{I}}-\Omega_{\mathrm{S}})-\frac{1}{2}\pi J&0\\0&0&0&-\frac{1}{2}(\Omega_{\mathrm{I}}+\Omega_{\mathrm{S}})+\frac{1}{2}\pi J\end{pmatrix} \tag{9.13}$$

which is equivalent to

$$\mathbf{H'_{IS}}=\Omega_{\mathrm{I}}\mathbf{I_z}+\Omega_{\mathrm{S}}\mathbf{S_z}+2\pi J\,\mathbf{I_zS_z}. \tag{9.14}$$

As will be seen later, this approximation is valid when $|2\pi J| \ll |\Omega_{\mathrm{I}}-\Omega_{\mathrm{S}}|$. With this simplification, it is clear that the eigenstates of the Hamiltonian are simply the basis states, $|\alpha_{\mathrm{I}}\alpha_{\mathrm{S}}\rangle$, $|\alpha_{\mathrm{I}}\beta_{\mathrm{S}}\rangle$, $|\beta_{\mathrm{I}}\alpha_{\mathrm{S}}\rangle$, $|\beta_{\mathrm{I}}\beta_{\mathrm{S}}\rangle$ and that the eigenvalues are the diagonal elements:

$$\begin{aligned}\varepsilon_1&=+\tfrac{1}{2}(\Omega_{\mathrm{I}}+\Omega_{\mathrm{S}})+\tfrac{1}{2}\pi J &\qquad \varepsilon_3&=-\tfrac{1}{2}(\Omega_{\mathrm{I}}-\Omega_{\mathrm{S}})-\tfrac{1}{2}\pi J\\ \varepsilon_2&=+\tfrac{1}{2}(\Omega_{\mathrm{I}}-\Omega_{\mathrm{S}})-\tfrac{1}{2}\pi J &\qquad \varepsilon_4&=-\tfrac{1}{2}(\Omega_{\mathrm{I}}+\Omega_{\mathrm{S}})+\tfrac{1}{2}\pi J.\end{aligned} \tag{9.15}$$

As discussed below, neglecting off-diagonal terms is equivalent to assuming that only the energy levels are changed by the coupling; the eigenstates remain unaffected. This corresponds to first-order perturbation theory. Comparison of Eqns 9.11 and 9.14 shows that it is the $\mathbf{I_xS_x}$ and $\mathbf{I_yS_y}$ terms that give rise to the discarded off-diagonal elements.

To calculate the free induction decay expected for a weakly coupled pair of homonuclear spins, we first consider the effect of a $90°_x$ pulse on the equilibrium state $\rho(0) = \mathbf{I_z} + \mathbf{S_z}$. The Hamiltonian during the pulse is (Appendix 8.4):

$$\mathbf{H} = \omega_1(\mathbf{I_x} + \mathbf{S_x})\,. \qquad (9.16)$$

Its exponential is easily obtained because $\mathbf{I_x}$ and $\mathbf{S_x}$ commute, so that

N.B. $e^{\mathbf{A+B}} = e^{\mathbf{A}}e^{\mathbf{B}}$ if $[\mathbf{A},\mathbf{B}] = 0$ (Appendix 8.2).

$$e^{\pm i\mathbf{H}t} = e^{\pm i\omega_1 t(\mathbf{I_x}+\mathbf{S_x})} = e^{\pm i\omega_1 t\mathbf{I_x}}e^{\pm i\omega_1 t\mathbf{S_x}}\,. \qquad (9.17)$$

Hence, using the single-spin 2×2 exponential matrix in Eqn 8.28, we have:

$$e^{\pm i\mathbf{H}t} = \begin{pmatrix} c & \pm is \\ \pm is & c \end{pmatrix} \otimes \begin{pmatrix} c & \pm is \\ \pm is & c \end{pmatrix} = \begin{pmatrix} c^2 & \pm isc & \pm isc & -s^2 \\ \pm isc & c^2 & -s^2 & \pm isc \\ \pm isc & -s^2 & c^2 & \pm isc \\ -s^2 & \pm isc & \pm isc & c^2 \end{pmatrix}, \qquad (9.18)$$

with

$$c = \cos(\tfrac{1}{2}\omega_1 t), \qquad s = \sin(\tfrac{1}{2}\omega_1 t). \qquad (9.19)$$

Matrix multiplication, using $\omega_1 t = \pi/2$ (i.e. $c = s = 1/\sqrt{2}$) gives

$$\rho(t) = e^{-i\mathbf{H}t}\rho(0)e^{+i\mathbf{H}t} = \begin{pmatrix} 0 & +\tfrac{1}{2}i & +\tfrac{1}{2}i & 0 \\ -\tfrac{1}{2}i & 0 & 0 & +\tfrac{1}{2}i \\ -\tfrac{1}{2}i & 0 & 0 & +\tfrac{1}{2}i \\ 0 & -\tfrac{1}{2}i & -\tfrac{1}{2}i & 0 \end{pmatrix} \equiv -(\mathbf{I_y} + \mathbf{S_y})\,. \qquad (9.20)$$

Thus a $90°_x$ pulse rotates z magnetization of both spins onto the $-y$ axis. Neither the resonance offsets nor the J coupling affects the excitation because of the assumption implicit in Eqn 9.16 that the radiofrequency field is the dominant interaction in the rotating frame.

The evolution during a subsequent period of free precession is easily calculated because the Hamiltonian is diagonal. Thus, in terms of the eigenvalues of Eqn 9.15:

$$e^{\pm i\mathbf{H}'_{\mathrm{IS}}t} = \begin{pmatrix} e^{\pm i\varepsilon_1 t} & 0 & 0 & 0 \\ 0 & e^{\pm i\varepsilon_2 t} & 0 & 0 \\ 0 & 0 & e^{\pm i\varepsilon_3 t} & 0 \\ 0 & 0 & 0 & e^{\pm i\varepsilon_4 t} \end{pmatrix} \qquad (9.21)$$

and

$$\rho(t) = e^{-i\mathbf{H}'_{IS}t}(-\mathbf{I}_y - \mathbf{S}_y)e^{+i\mathbf{H}'_{IS}t}$$

$$= \begin{pmatrix} 0 & \frac{i}{2}e^{i(\varepsilon_2-\varepsilon_1)t} & \frac{i}{2}e^{i(\varepsilon_3-\varepsilon_1)t} & 0 \\ -\frac{i}{2}e^{-i(\varepsilon_2-\varepsilon_1)t} & 0 & 0 & \frac{i}{2}e^{i(\varepsilon_4-\varepsilon_2)t} \\ -\frac{i}{2}e^{-i(\varepsilon_3-\varepsilon_1)t} & 0 & 0 & \frac{i}{2}e^{i(\varepsilon_4-\varepsilon_3)t} \\ 0 & -\frac{i}{2}e^{-i(\varepsilon_4-\varepsilon_2)t} & -\frac{i}{2}e^{-i(\varepsilon_4-\varepsilon_3)t} & 0 \end{pmatrix}. \quad (9.22)$$

The expectation values of the observable in-phase x and y magnetization for spin I are therefore:

$$\begin{aligned} \langle \mathbf{I}_y \rangle &= \mathrm{Tr}\left[\rho(t)\mathbf{I}_y\right] = -\tfrac{1}{2}\cos[(\Omega_I + \pi J)t] - \tfrac{1}{2}\cos[(\Omega_I - \pi J)t] \\ &= -\cos\Omega_I t \cos \pi J t \\ \langle \mathbf{I}_x \rangle &= \mathrm{Tr}\left[\rho(t)\mathbf{I}_x\right] = +\tfrac{1}{2}\sin[(\Omega_I + \pi J)t] + \tfrac{1}{2}\sin[(\Omega_I - \pi J)t] \\ &= +\sin\Omega_I t \cos \pi J t \end{aligned} \quad (9.23)$$

and for the I-spin antiphase operators:

$$\begin{aligned} \langle 2\mathbf{I}_y\mathbf{S}_z \rangle &= 2\mathrm{Tr}\left[\rho(t)\mathbf{I}_y\mathbf{S}_z\right] = -\tfrac{1}{2}\cos[(\Omega_I + \pi J)t] + \tfrac{1}{2}\cos[(\Omega_I - \pi J)t] \\ &= +\sin\Omega_I t \sin \pi J t \\ \langle 2\mathbf{I}_x\mathbf{S}_z \rangle &= 2\mathrm{Tr}\left[\rho(t)\mathbf{I}_x\mathbf{S}_z\right] = +\tfrac{1}{2}\sin[(\Omega_I + \pi J)t] - \tfrac{1}{2}\sin[(\Omega_I - \pi J)t] \\ &= +\cos\Omega_I t \sin \pi J t \end{aligned} \quad (9.24)$$

in agreement with Eqn 3.10. The frequencies $\Omega_I \pm \pi J$ correspond to the two components of the I-spin doublet. The signs in front of the $\cos[(\Omega_I \pm \pi J)t]$ and $\sin[(\Omega_I \pm \pi J)t]$ terms indicate in-phase and antiphase doublets. Similar expressions can, of course, be obtained for the S spins.

9.5 Weak coupling: a more cunning approach

Alternatively, and with *much less effort*, we can use the approach introduced in Section 8.6 and summarized in Eqn 8.41:

$$\begin{aligned} &\mathbf{A} \xrightarrow{b\mathbf{B}t} \mathbf{A}\cos bt - \mathbf{C}\sin bt \\ &[\mathbf{A},\mathbf{B}] = i\mathbf{C} \\ &[\mathbf{B},\mathbf{C}] = i\mathbf{A}. \end{aligned} \quad (9.25)$$

Remember that operators and their matrix representations are used interchangeably.

We first note that the three parts of the Hamiltonian, Eqn 9.14, all commute with each other. This allows us to calculate their effects sequentially and in any order (see Appendix 8.2).

This should be recognized as one of the properties of product operators that make them so convenient.

During the period of free precession following the 90° pulse, the evolution under the J coupling is described by the operators: $\mathbf{A} = -(\mathbf{I}_y + \mathbf{S}_y)$, $\mathbf{B} = 2\mathbf{I}_z\mathbf{S}_z$, $\mathbf{C} = -(2\mathbf{I}_x\mathbf{S}_z + 2\mathbf{I}_z\mathbf{S}_x)$, with $b = \pi J$ so that

$$-(\mathbf{I}_y + \mathbf{S}_y) \longrightarrow -(\mathbf{I}_y + \mathbf{S}_y)\cos \pi J t + (2\mathbf{I}_x\mathbf{S}_z + 2\mathbf{I}_z\mathbf{S}_x)\sin \pi J t\,, \quad (9.26)$$

The commutators (Appendix 9.3) are: $[\mathbf{A}, \mathbf{B}] = -i(2\mathbf{I}_x\mathbf{S}_z + 2\mathbf{I}_z\mathbf{S}_x) = i\mathbf{C}$, $[\mathbf{B},\mathbf{C}] = -i(\mathbf{I}_y + \mathbf{S}_y) = i\mathbf{A}$.

One can then repeat the process for each of the Zeeman interactions in turn, just as one would do with product operators, to obtain the results in Eqns 9.23 and 9.24.

thus explaining the origin of the antiphase product operators, e.g. $2I_xS_z$, the factor of two in front of them, and the fact that the evolution under the J coupling occurs at frequency πJ (Eqn 3.8). As in the case of isolated spins in Chapter 8, we see the correspondence of product operators and density matrices.

9.6 Equivalent spins

Spins are said to be equivalent if they have identical chemical shifts and J couplings.

Many molecules contain groups of two or more equivalent nuclei, such as the three protons in a methyl group. Even though such nuclei are coupled to one another, no splitting is observed in the NMR spectrum. We can now see why.

The brute-force approach would be to write down the Hamiltonian (Eqn 9.12) with $\Omega_I = \Omega_S = \Omega$:

$$\mathbf{H}_{\mathbf{IS}} = \Omega(\mathbf{I_z} + \mathbf{S_z}) + 2\pi J\,\mathbf{I}\cdot\mathbf{S} = \begin{pmatrix} \Omega + \frac{1}{2}\pi J & 0 & 0 & 0 \\ 0 & -\frac{1}{2}\pi J & \pi J & 0 \\ 0 & \pi J & -\frac{1}{2}\pi J & 0 \\ 0 & 0 & 0 & -\Omega + \frac{1}{2}\pi J \end{pmatrix} \tag{9.27}$$

and calculate its exponential

$$e^{\pm i\mathbf{H}_{\mathbf{IS}}t}$$

$$= \begin{pmatrix} e^{\pm i(\Omega+\pi J/2)t} & 0 & 0 & 0 \\ 0 & \frac{1}{2}e^{\pm i\pi Jt/2} + \frac{1}{2}e^{\mp 3i\pi Jt/2} & \frac{1}{2}e^{\pm i\pi Jt/2} - \frac{1}{2}e^{\mp 3i\pi Jt/2} & 0 \\ 0 & \frac{1}{2}e^{\pm i\pi Jt/2} - \frac{1}{2}e^{\mp 3i\pi Jt/2} & \frac{1}{2}e^{\pm i\pi Jt/2} + \frac{1}{2}e^{\mp 3i\pi Jt/2} & 0 \\ 0 & 0 & 0 & e^{\pm i(-\Omega+\pi J/2)t} \end{pmatrix} \tag{9.28}$$

using the method described in Appendix 8.2, and then to show by matrix multiplication that the density matrix during the free induction decay is

$$\rho(t) = e^{-i\mathbf{H}_{\mathbf{IS}}t}(-\mathbf{I_y} - \mathbf{S_y})e^{+i\mathbf{H}_{\mathbf{IS}}t} = \frac{i}{2}\begin{pmatrix} 0 & e^{-i\Omega t} & e^{-i\Omega t} & 0 \\ -e^{+i\Omega t} & 0 & 0 & e^{-i\Omega t} \\ -e^{+i\Omega t} & 0 & 0 & e^{-i\Omega t} \\ 0 & -e^{+i\Omega t} & -e^{+i\Omega t} & 0 \end{pmatrix} \tag{9.29}$$

and hence

We evaluate the total x and y magnetization because I and S are indistinguishable.

$$\begin{aligned} \langle \mathbf{I_y} + \mathbf{S_y} \rangle &= \mathrm{Tr}\left[\rho(t)(\mathbf{I_y} + \mathbf{S_y})\right] = -2\cos\Omega t \\ \langle \mathbf{I_x} + \mathbf{S_x} \rangle &= \mathrm{Tr}\left[\rho(t)(\mathbf{I_x} + \mathbf{S_x})\right] = +2\sin\Omega t. \end{aligned} \tag{9.30}$$

Thus, the spectrum will comprise a single line at the chemical shift frequency Ω and *no* splitting due to J coupling.

Fortunately, there is a more elegant and general method. Consider a spin system described by a Hamiltonian **H** comprising two parts with the following properties:

$$\mathbf{H} = \mathbf{H_1} + \mathbf{H_2} \quad \text{and} \quad \left[\mathbf{H_1}, \mathbf{H_2}\right] = \left[\mathbf{F_x}, \mathbf{H_2}\right] = \left[\mathbf{F_y}, \mathbf{H_2}\right] = 0 \tag{9.31}$$

where $\mathbf{F_x}$ and $\mathbf{F_y}$ are the total x and y operators, summed over all spins. The observable x magnetization is given by the Liouville–von Neumann equation as

e.g. $\mathbf{F_x} = \mathbf{I_x} + \mathbf{S_x}$ for a two-spin system.

$$\begin{aligned}
\mathrm{Tr}\left[\mathbf{F_x}\mathrm{e}^{-i\mathbf{H}t}\rho(0)\mathrm{e}^{+i\mathbf{H}t}\right] &= \mathrm{Tr}\left[\mathbf{F_x}\mathrm{e}^{-i\mathbf{H}_2t}\mathrm{e}^{-i\mathbf{H}_1t}\rho(0)\mathrm{e}^{+i\mathbf{H}_1t}\mathrm{e}^{+i\mathbf{H}_2t}\right] \\
&= \mathrm{Tr}\left[\mathrm{e}^{+i\mathbf{H}_2t}\mathbf{F_x}\mathrm{e}^{-i\mathbf{H}_2t}\mathrm{e}^{-i\mathbf{H}_1t}\rho(0)\mathrm{e}^{+i\mathbf{H}_1t}\right] \\
&= \mathrm{Tr}\left[\mathbf{F_x}\mathrm{e}^{+i\mathbf{H}_2t}\mathrm{e}^{-i\mathbf{H}_2t}\mathrm{e}^{-i\mathbf{H}_1t}\rho(0)\mathrm{e}^{+i\mathbf{H}_1t}\right] \\
&= \mathrm{Tr}\left[\mathbf{F_x}\mathrm{e}^{-i\mathbf{H}_1t}\rho(0)\mathrm{e}^{+i\mathbf{H}_1t}\right].
\end{aligned} \tag{9.32}$$

These expressions require some of the properties of exponential matrices outlined in Appendix 8.2. In going from line 1 to line 2 we have used the fact that the trace is invariant to cyclic permutation of the operators within it (Appendix 7.1).

Thus, $\mathbf{H_2}$ has no observable effect, whatever the initial state $\rho(0)$. The same holds for the y component of magnetization.

To see how all this relates to equivalence, consider a system of four spins, $\mathrm{I_1}$, $\mathrm{I_2}$, S and R in which $\mathrm{I_1}$ and $\mathrm{I_2}$ are equivalent. $\mathbf{H_2}$ is the coupling term between the I spins:

$$\mathbf{H_2} = 2\pi J_{\mathrm{I_1I_2}}\mathbf{I_1}\cdot\mathbf{I_2} \tag{9.33}$$

and it commutes with all the other terms in the Hamiltonian:

$$\begin{aligned}
\mathbf{H_1} = {}& \Omega_{\mathrm{I}}(\mathbf{I_{1z}} + \mathbf{I_{2z}}) + \Omega_{\mathrm{S}}\mathbf{S_z} + \Omega_{\mathrm{R}}\mathbf{R_z} + 2\pi J_{\mathrm{IS}}(\mathbf{I_1} + \mathbf{I_2})\cdot\mathbf{S} \\
&+ 2\pi J_{\mathrm{IR}}(\mathbf{I_1} + \mathbf{I_2})\cdot\mathbf{R} + 2\pi J_{\mathrm{SR}}\mathbf{S}\cdot\mathbf{R}.
\end{aligned} \tag{9.34}$$

The IS subscript on J_{IS} is briefly reinstated here to avoid confusion.

$\mathbf{H_2}$ also commutes with the total x and y operators, e.g. $\mathbf{F_x} = \mathbf{I_{1x}} + \mathbf{I_{2x}} + \mathbf{S_x} + \mathbf{R_x}$, and therefore has no effect on the observed signal. The same result is *not* true for inequivalent spins because the coupling term does not commute with the Zeeman interactions of the I spins:

$$[2\pi J_{\mathrm{I_1I_2}}\mathbf{I_1}\cdot\mathbf{I_2}, (\Omega_{\mathrm{I_1}}\mathbf{I_{1z}} + \Omega_{\mathrm{I_2}}\mathbf{I_{2z}})] \neq \mathbf{0}. \tag{9.35}$$

This approach is easily generalized to more than two equivalent spins and more than one group of equivalent spins.

9.7 Evolution of multiple-quantum coherence

The Liouville–von Neumann equation allows us to understand one of the properties of product operators stated without proof in Part A, namely that the zero- and double-quantum coherences of spins I and S do *not* evolve under the J_{IS} coupling (Section 4.3). It is clear from Eqn 8.11 that if the initial density operator and the Hamiltonian commute, then there can be no spin evolution. It is easily verified that all four of the operators involved in these coherences ($2\mathbf{I_xS_x}$, $2\mathbf{I_xS_y}$, $2\mathbf{I_yS_x}$ and $2\mathbf{I_yS_y}$, Eqn 4.5) commute with the weak coupling Hamiltonian $2\pi J_{\mathrm{IS}}\mathbf{I_zS_z}$. This may be done by multiplying the matrix representations of the operators (Appendix 9.2), or by using Eqn 7.29 and its cyclic permutations. For example:

$$
\begin{aligned}
\left[\mathbf{I_xS_y},\mathbf{I_zS_z}\right] &= \mathbf{I_xS_yI_zS_z} - \mathbf{I_zS_zI_xS_y} = \mathbf{I_xI_zS_yS_z} - \mathbf{I_zI_xS_zS_y} \\
&= \left(-\tfrac{1}{2}i\mathbf{I_y}\right)\left(\tfrac{1}{2}i\mathbf{S_x}\right) - \left(\tfrac{1}{2}i\mathbf{I_y}\right)\left(-\tfrac{1}{2}i\mathbf{S_x}\right) = \tfrac{1}{4}\mathbf{I_yS_x} - \tfrac{1}{4}\mathbf{I_yS_x} \quad (9.36) \\
&= 0,
\end{aligned}
$$

where we have the fact that all I operators commute with all S operators. Appendix 9.3 gives a full list of the commutation relations for a two-spin system.

9.8 TOCSY

We are at last in a position to justify the expressions given in Eqn 5.13 for the evolution during a period of spin locking, as in the TOCSY experiment (Section 5.5). The Hamiltonian in the rotating frame for an IS spin system, with a spin-locking field along the x axis, is

$$\hat{H} = \omega_1^{\mathrm{I}}\hat{I}_x + \omega_1^{\mathrm{S}}\hat{S}_x + \Omega_{\mathrm{I}}\hat{I}_z + \Omega_{\mathrm{S}}\hat{S}_z + 2\pi J\hat{I}\cdot\hat{S} \qquad (9.37)$$

where $\omega_1{}^{\mathrm{I}}/\gamma_{\mathrm{I}}$ and $\omega_1{}^{\mathrm{S}}/\gamma_{\mathrm{S}}$ are the radiofrequency field strengths experienced by the two spins. For strong spin-locking fields ($\omega_1{}^{\mathrm{I}} >> \Omega_{\mathrm{I}}$, $\omega_1{}^{\mathrm{S}} >> \Omega_{\mathrm{S}}$), both spins experience an effective field along the x axis, so that the Zeeman terms in Eqn 9.36 can be dropped. In a homonuclear spin system, such as a pair of ^{1}H nuclei, $\omega_1{}^{\mathrm{I}}$ and $\omega_1{}^{\mathrm{S}}$ are almost identical; they differ by a few parts per million, but since $\omega_1/2\pi$ is typically about 10 kHz this difference will be only a tiny fraction of a Hz. Thus,

$$\hat{H} = \omega_1\left(\hat{I}_x + \hat{S}_x\right) + 2\pi J\hat{I}\cdot\hat{S}, \qquad (9.38)$$

which can be seen to have exactly the same form as the Hamiltonian for two equivalent spins (Eqn 9.27) except that Ω is replaced by ω_1, and z by x. In a heteronuclear spin system different radiofrequency fields are used to irradiate the two spins, so that $\omega_1{}^{\mathrm{I}}$ and $\omega_1{}^{\mathrm{S}}$ can be separately controlled. In particular, the two frequencies can be chosen to be the same (the Hartmann–Hahn condition), once again making the spins effectively equivalent.

Since the two parts of Eqn 9.38 commute, we can treat them sequentially and in any order. The first term has no effect on either $\mathbf{I_x}$ or $\mathbf{S_x}$ because they both commute with $(\mathbf{I_x} + \mathbf{S_x})$. It is the J coupling that produces the desired effect. Using Eqn 9.25 with $\mathbf{A} = (\mathbf{I_x} - \mathbf{S_x})$, $\mathbf{B} = \mathbf{I}\cdot\mathbf{S}$, $\mathbf{C} = -2\mathbf{I_yS_z} + 2\mathbf{I_zS_y}$, and $b = 2\pi J$, one finds (cf. Eqn 5.13)

$$\mathbf{I_x} - \mathbf{S_x} \rightarrow (\mathbf{I_x} - \mathbf{S_x})\cos 2\pi Jt + (2\mathbf{I_yS_z} - 2\mathbf{I_zS_y})\sin 2\pi Jt\,. \qquad (9.39)$$

$\mathbf{I}\cdot\mathbf{S}$ does, however, commute with $\mathbf{I_x} + \mathbf{S_x}$, so that the sum of the x operators does not evolve.

Similar effects follow for $\mathbf{I_y} \pm \mathbf{S_y}$ and $\mathbf{I_z} \pm \mathbf{S_z}$. The y and z magnetizations also evolve under the first term in Eqn 9.38, e.g.

$$\mathbf{I_y} \rightarrow \mathbf{I_y}\cos\omega_1 t + \mathbf{I_z}\sin\omega_1 t\,. \qquad (9.40)$$

For genuinely equivalent spins the evolution described by Eqn 9.39 can have no discernible effect, since the *observable* magnetization does not evolve under the J coupling (Section 9.6). Spins which are rendered temporarily equivalent, however, can show much more complex behaviour, as they can enter the spin-lock period in a variety of initial states, and can be observed as inequivalent spins at the end of the spin-locking period.

Appendix 9.1 Direct products

Direct products, also called tensor or outer products, provide a convenient way of describing a system made up of two or more subsystems. If each of two subsystems is described by an $n \times n$ matrix, then the whole system requires an $n^2 \times n^2$ matrix to describe it completely. This matrix may be obtained as follows: each element in the first matrix is multiplied by the second matrix, and the inner brackets are deleted:

$$\begin{pmatrix} a & b \\ c & d \end{pmatrix} \otimes \begin{pmatrix} \alpha & \beta \\ \gamma & \delta \end{pmatrix} \rightarrow \begin{pmatrix} a\begin{pmatrix} \alpha & \beta \\ \gamma & \delta \end{pmatrix} & b\begin{pmatrix} \alpha & \beta \\ \gamma & \delta \end{pmatrix} \\ c\begin{pmatrix} \alpha & \beta \\ \gamma & \delta \end{pmatrix} & d\begin{pmatrix} \alpha & \beta \\ \gamma & \delta \end{pmatrix} \end{pmatrix} = \begin{pmatrix} a\alpha & a\beta & b\alpha & b\beta \\ a\gamma & a\delta & b\gamma & b\delta \\ c\alpha & c\beta & d\alpha & d\beta \\ c\gamma & c\delta & d\gamma & d\delta \end{pmatrix}. \quad (9.41)$$

Note that the order in which two matrices are multiplied together has no fundamental significance, but a consistent order must be used throughout a calculation; this corresponds to a consistent labelling of the two subsystems, I and S.

Appendix 9.2 Matrix representations of two-spin operators

$$\mathbf{I_x} = \begin{pmatrix} 0 & 0 & \frac{1}{2} & 0 \\ 0 & 0 & 0 & \frac{1}{2} \\ \frac{1}{2} & 0 & 0 & 0 \\ 0 & \frac{1}{2} & 0 & 0 \end{pmatrix} \qquad \mathbf{I_y} = \begin{pmatrix} 0 & 0 & -\frac{1}{2}i & 0 \\ 0 & 0 & 0 & -\frac{1}{2}i \\ \frac{1}{2}i & 0 & 0 & 0 \\ 0 & \frac{1}{2}i & 0 & 0 \end{pmatrix} \qquad \mathbf{I_z} = \begin{pmatrix} \frac{1}{2} & 0 & 0 & 0 \\ 0 & \frac{1}{2} & 0 & 0 \\ 0 & 0 & -\frac{1}{2} & 0 \\ 0 & 0 & 0 & -\frac{1}{2} \end{pmatrix}$$

$$\mathbf{S_x} = \begin{pmatrix} 0 & \frac{1}{2} & 0 & 0 \\ \frac{1}{2} & 0 & 0 & 0 \\ 0 & 0 & 0 & \frac{1}{2} \\ 0 & 0 & \frac{1}{2} & 0 \end{pmatrix} \qquad \mathbf{S_y} = \begin{pmatrix} 0 & -\frac{1}{2}i & 0 & 0 \\ \frac{1}{2}i & 0 & 0 & 0 \\ 0 & 0 & 0 & -\frac{1}{2}i \\ 0 & 0 & \frac{1}{2}i & 0 \end{pmatrix} \qquad \mathbf{S_z} = \begin{pmatrix} \frac{1}{2} & 0 & 0 & 0 \\ 0 & -\frac{1}{2} & 0 & 0 \\ 0 & 0 & \frac{1}{2} & 0 \\ 0 & 0 & 0 & -\frac{1}{2} \end{pmatrix}$$

$$2\mathbf{I_x S_x} = \begin{pmatrix} 0 & 0 & 0 & \frac{1}{2} \\ 0 & 0 & \frac{1}{2} & 0 \\ 0 & \frac{1}{2} & 0 & 0 \\ \frac{1}{2} & 0 & 0 & 0 \end{pmatrix} \qquad 2\mathbf{I_x S_y} = \begin{pmatrix} 0 & 0 & 0 & -\frac{1}{2}i \\ 0 & 0 & \frac{1}{2}i & 0 \\ 0 & -\frac{1}{2}i & 0 & 0 \\ \frac{1}{2}i & 0 & 0 & 0 \end{pmatrix} \qquad 2\mathbf{I_x S_z} = \begin{pmatrix} 0 & 0 & \frac{1}{2} & 0 \\ 0 & 0 & 0 & -\frac{1}{2} \\ \frac{1}{2} & 0 & 0 & 0 \\ 0 & -\frac{1}{2} & 0 & 0 \end{pmatrix}$$

$$2\mathbf{I_y S_x} = \begin{pmatrix} 0 & 0 & 0 & -\frac{1}{2}i \\ 0 & 0 & -\frac{1}{2}i & 0 \\ 0 & \frac{1}{2}i & 0 & 0 \\ \frac{1}{2}i & 0 & 0 & 0 \end{pmatrix} \qquad 2\mathbf{I_y S_y} = \begin{pmatrix} 0 & 0 & 0 & -\frac{1}{2} \\ 0 & 0 & \frac{1}{2} & 0 \\ 0 & \frac{1}{2} & 0 & 0 \\ -\frac{1}{2} & 0 & 0 & 0 \end{pmatrix} \qquad 2\mathbf{I_y S_z} = \begin{pmatrix} 0 & 0 & -\frac{1}{2}i & 0 \\ 0 & 0 & 0 & \frac{1}{2}i \\ \frac{1}{2}i & 0 & 0 & 0 \\ 0 & -\frac{1}{2}i & 0 & 0 \end{pmatrix}$$

$$2\mathbf{I_z S_x} = \begin{pmatrix} 0 & \frac{1}{2} & 0 & 0 \\ \frac{1}{2} & 0 & 0 & 0 \\ 0 & 0 & 0 & -\frac{1}{2} \\ 0 & 0 & -\frac{1}{2} & 0 \end{pmatrix} \qquad 2\mathbf{I_z S_y} = \begin{pmatrix} 0 & -\frac{1}{2}i & 0 & 0 \\ \frac{1}{2}i & 0 & 0 & 0 \\ 0 & 0 & 0 & \frac{1}{2}i \\ 0 & 0 & -\frac{1}{2}i & 0 \end{pmatrix} \qquad 2\mathbf{I_z S_z} = \begin{pmatrix} \frac{1}{2} & 0 & 0 & 0 \\ 0 & -\frac{1}{2} & 0 & 0 \\ 0 & 0 & -\frac{1}{2} & 0 \\ 0 & 0 & 0 & \frac{1}{2} \end{pmatrix}$$

$$\mathbf{ZQ_x} = \begin{pmatrix} 0 & 0 & 0 & 0 \\ 0 & 0 & \frac{1}{2} & 0 \\ 0 & \frac{1}{2} & 0 & 0 \\ 0 & 0 & 0 & 0 \end{pmatrix} \qquad \mathbf{ZQ_y} = \begin{pmatrix} 0 & 0 & 0 & 0 \\ 0 & 0 & -\frac{1}{2}i & 0 \\ 0 & \frac{1}{2}i & 0 & 0 \\ 0 & 0 & 0 & 0 \end{pmatrix} \qquad \mathbf{I \cdot S} = \begin{pmatrix} \frac{1}{4} & 0 & 0 & 0 \\ 0 & -\frac{1}{4} & \frac{1}{2} & 0 \\ 0 & \frac{1}{2} & -\frac{1}{4} & 0 \\ 0 & 0 & 0 & \frac{1}{4} \end{pmatrix}$$

$$\mathbf{DQ_x} = \begin{pmatrix} 0 & 0 & 0 & \frac{1}{2} \\ 0 & 0 & 0 & 0 \\ 0 & 0 & 0 & 0 \\ \frac{1}{2} & 0 & 0 & 0 \end{pmatrix} \qquad \mathbf{DQ_y} = \begin{pmatrix} 0 & 0 & 0 & -\frac{1}{2}i \\ 0 & 0 & 0 & 0 \\ 0 & 0 & 0 & 0 \\ \frac{1}{2}i & 0 & 0 & 0 \end{pmatrix}$$

Appendix 9.3 Operator commutators

	I_x	I_y	I_z	S_x	S_y	S_z	$2I_xS_x$	$2I_xS_y$	$2I_xS_z$	$2I_yS_x$	$2I_yS_y$	$2I_yS_z$	$2I_zS_x$	$2I_zS_y$	$2I_zS_z$
I_x	0	$+iI_z$	$-iI_y$	0	0	0	0	0	0	$+i2I_zS_x$	$+i2I_zS_y$	$+i2I_zS_z$	$-i2I_yS_x$	$-i2I_yS_y$	$-i2I_yS_z$
I_y	$-iI_z$	0	$+iI_x$	0	0	0	$-i2I_zS_x$	$-i2I_zS_y$	$-i2I_zS_z$	0	0	0	$+i2I_xS_x$	$+i2I_xS_y$	$+i2I_xS_z$
I_z	$+iI_y$	$-iI_x$	0	0	0	0	$+i2I_yS_x$	$+i2I_yS_y$	$+i2I_yS_z$	$-i2I_xS_x$	$-i2I_xS_y$	$-i2I_xS_z$	0	0	0
S_x	0	0	0	0	$+iS_z$	$-iS_y$	0	$+i2I_xS_z$	$-2iI_xS_y$	0	$+i2I_yS_z$	$-i2I_yS_y$	0	$+i2I_zS_z$	$-i2I_zS_y$
S_y	0	0	0	$-iS_z$	0	$+iS_x$	$-i2I_xS_z$	0	$+i2I_xS_x$	$-i2I_yS_z$	0	$+i2I_yS_x$	$-i2I_zS_z$	0	$+i2I_zS_x$
S_z	0	0	0	$+iS_y$	$-iS_x$	0	$+i2I_xS_y$	$-i2I_xS_x$	0	$+i2I_yS_y$	$-i2I_yS_x$	0	$+i2I_zS_y$	$-i2I_zS_x$	0
$2I_xS_x$	0	$+i2I_zS_x$	$-i2I_yS_x$	0	$+i2I_xS_z$	$-i2I_xS_y$	0	$+iS_z$	$-iS_y$	$+iI_z$	0	0	$-iI_y$	0	0
$2I_xS_y$	0	$+i2I_zS_y$	$-i2I_yS_y$	$-i2I_xS_z$	0	$+i2I_xS_x$	$-iS_z$	0	$+iS_x$	0	$+iI_z$	0	0	$-iI_y$	0
$2I_xS_z$	0	$+i2I_zS_z$	$-i2I_yS_z$	$+2iI_xS_y$	$-i2I_xS_x$	0	$+iS_y$	$-iS_x$	0	0	0	$+iI_z$	0	0	$-iI_y$
$2I_yS_x$	$-i2I_zS_x$	0	$+i2I_xS_x$	0	$+i2I_yS_z$	$-i2I_yS_y$	$-iI_z$	0	0	0	$+iS_z$	$-iS_y$	$+iI_x$	0	0
$2I_yS_y$	$-i2I_zS_y$	0	$+i2I_xS_y$	$-i2I_yS_z$	0	$+i2I_yS_x$	0	$-iI_z$	0	$-iS_z$	0	$+iS_x$	0	$+iI_x$	0
$2I_yS_z$	$-i2I_zS_z$	0	$+i2I_xS_z$	$+i2I_yS_y$	$-i2I_yS_x$	0	0	0	$-iI_z$	$+iS_y$	$-iS_x$	0	0	0	$+iI_x$
$2I_zS_x$	$+i2I_yS_x$	$-i2I_xS_x$	0	0	$+i2I_zS_z$	$-i2I_zS_y$	$+iI_y$	0	0	$-iI_x$	0	0	0	$+iS_z$	$-iS_y$
$2I_zS_y$	$+i2I_yS_y$	$-i2I_xS_y$	0	$-i2I_zS_z$	0	$+i2I_zS_x$	0	$+iI_y$	0	0	$-iI_x$	0	$-iS_z$	0	$+iS_x$
$2I_zS_z$	$+i2I_yS_z$	$-i2I_xS_z$	0	$+i2I_zS_y$	$-i2I_zS_x$	0	0	0	$+iI_y$	0	0	$-iI_x$	$+iS_y$	$-iS_x$	0

The operator hats (^) have been omitted here for simplicity.

10 Strong coupling

10.1 Introduction

Spin systems comprising weakly coupled and/or equivalent spins are straightforward to describe quantum mechanically, as we saw in Chapter 9, because the evolution induced by the various interactions can be calculated sequentially, and in any order. This is no longer true for strong coupling because the coupling term $2\pi J\mathbf{I}{\cdot}\mathbf{S}$ does not commute with the other parts of the spin Hamiltonian. For this reason, product operators are useless and there is no option but to use the methods in Chapters 8 and 9.

We start by calculating the free induction decay for a pair of strongly coupled spin-1/2 nuclei excited by a 90° pulse, and then extend the treatment to a spin echo. These calculations illustrate how one might approach more complicated problems.

10.2 Free induction decay

The full Hamiltonian $\mathbf{H_{IS}}$ for a pair of J-coupled spins I and S has been given in Eqns 9.11 and 9.12. Its eigenvalues and eigenvectors are easily obtained (Appendix 8.2):

$$\begin{aligned} \varepsilon_1/2\pi &= +\nu + \tfrac{1}{4}J \qquad & \varepsilon_3/2\pi &= -\tfrac{1}{2}\varepsilon - \tfrac{1}{4}J \\ \varepsilon_2/2\pi &= +\tfrac{1}{2}\varepsilon - \tfrac{1}{4}J \qquad & \varepsilon_4/2\pi &= -\nu + \tfrac{1}{4}J \end{aligned} \tag{10.1}$$

These expressions may be verified by checking that, e.g. $\hat{H}\,|\psi_2\rangle = \varepsilon_2|\psi_2\rangle$.

$$\begin{aligned} |\psi_1\rangle &= |\alpha_I\alpha_S\rangle \\ |\psi_2\rangle &= \cos\theta|\alpha_I\beta_S\rangle + \sin\theta|\beta_I\alpha_S\rangle \\ |\psi_3\rangle &= -\sin\theta|\alpha_I\beta_S\rangle + \cos\theta|\beta_I\alpha_S\rangle \\ |\psi_4\rangle &= |\beta_I\beta_S\rangle \end{aligned} \tag{10.2}$$

where

$$2\pi\nu = \tfrac{1}{2}(\Omega_I + \Omega_S), \qquad \varepsilon = \sqrt{J^2 + \delta^2}, \qquad 2\pi\delta = \Omega_I - \Omega_S, \tag{10.3}$$

and the *mixing angle* θ is given by

$$\tan 2\theta = \frac{J}{\delta}, \quad \text{or} \quad J = \varepsilon\sin 2\theta \quad \text{and} \quad \delta = \varepsilon\cos 2\theta\,. \tag{10.4}$$

When $|J| << |\delta|$, θ is zero, the eigenstates reduce to those for weak coupling, i.e. $|\alpha_I\alpha_S\rangle$, $|\alpha_I\beta_S\rangle$, $|\beta_I\alpha_S\rangle$, $|\beta_I\beta_S\rangle$, with eigenvalues as in Eqn 9.15.

We calculate first the y component of the free induction decay following a $90°_x$ pulse

$$\langle \hat{F}_y \rangle = \mathrm{Tr}\left[\mathbf{F_y}\rho(t)\right], \tag{10.5}$$

where $\mathbf{F_y}$ is the matrix representation of the operator for the total y magnetization

$$\mathbf{F_y} = \mathbf{I_y} + \mathbf{S_y} \tag{10.6}$$

and the density operator is given by

$$\rho(t) = \mathrm{e}^{-i\mathbf{H}_{\mathbf{IS}}t}\mathrm{e}^{-i\frac{1}{2}\pi\mathbf{F_x}}\rho(0)\mathrm{e}^{+i\frac{1}{2}\pi\mathbf{F_x}}\mathrm{e}^{+i\mathbf{H}_{\mathbf{IS}}t} \tag{10.7}$$

cf. Eqn 8.16.

with initial state

$$\rho(0) = \mathbf{I_z} + \mathbf{S_z} = \mathbf{F_z}. \tag{10.8}$$

The exponential operators for the period after the pulse are given by

$$\mathrm{e}^{\pm i\mathbf{H}_{\mathbf{IS}}t} = \mathbf{V}\mathrm{e}^{\pm i\mathbf{h}t}\mathbf{V}^{\mathsf{T}} \tag{10.9}$$

cf. Appendix 8.2. Note that $\mathbf{V}^{-1} = \mathbf{V}^{\mathsf{T}}$ here.

in which $\mathbf{h}$ is a diagonal matrix with the eigenvalues of $\mathbf{H}_{\mathbf{IS}}$ as its elements, and $\exp(\pm i\mathbf{h}t)$ is given by Eqn 9.21 using the eigenvalues in Eqn 10.1. $\mathbf{V}$ is the matrix of eigenvectors (as columns):

$$\mathbf{V} = \begin{pmatrix} 1 & 0 & 0 & 0 \\ 0 & c & -s & 0 \\ 0 & s & c & 0 \\ 0 & 0 & 0 & 1 \end{pmatrix} \quad \text{with} \quad \begin{aligned} s &= \sin\theta \\ c &= \cos\theta, \end{aligned} \tag{10.10}$$

Putting everything together gives:

$$\langle \hat{F}_y \rangle = \mathrm{Tr}\left[\mathbf{F_y}\mathbf{V}\mathrm{e}^{-i\mathbf{h}t}\mathbf{V}^{\mathsf{T}}\mathrm{e}^{-i\frac{1}{2}\pi\mathbf{F_x}}\rho(0)\mathrm{e}^{+i\frac{1}{2}\pi\mathbf{F_x}}\mathbf{V}\mathrm{e}^{+i\mathbf{h}t}\mathbf{V}^{\mathsf{T}}\right]. \tag{10.11}$$

Cyclically permuting $\mathbf{V}^{\mathsf{T}}$ within the trace, gives:

cf. Appendix 7.1.

$$\langle \hat{F}_y \rangle = \mathrm{Tr}\left[\left\{\mathbf{V}^{\mathsf{T}}\mathbf{F_y}\mathbf{V}\right\}\mathrm{e}^{-i\mathbf{h}t}\left\{\mathbf{V}^{\mathsf{T}}\left(\mathrm{e}^{-i\frac{1}{2}\pi\mathbf{F_x}}\rho(0)\mathrm{e}^{+i\frac{1}{2}\pi\mathbf{F_x}}\right)\mathbf{V}\right\}\mathrm{e}^{+i\mathbf{h}t}\right]. \tag{10.12}$$

All we have to do now is to multiply the matrices together and take the trace. This task is less daunting than it may seem when the matrices are grouped as indicated. The part which gives the effect of the 90° pulse, is already known (Eqn 9.20)

$$\mathrm{e}^{-i\frac{1}{2}\pi\mathbf{F_x}}\rho(0)\mathrm{e}^{+i\frac{1}{2}\pi\mathbf{F_x}} = \mathrm{e}^{-i\frac{1}{2}\pi\mathbf{F_x}}\mathbf{F_z}\mathrm{e}^{+i\frac{1}{2}\pi\mathbf{F_x}} = -\mathbf{F_y} \tag{10.13}$$

so that

$$\langle \hat{F}_y \rangle = -\mathrm{Tr}\left[\left\{\mathbf{V}^{\mathsf{T}}\mathbf{F_y}\mathbf{V}\right\}\mathrm{e}^{-i\mathbf{h}t}\left\{\mathbf{V}^{\mathsf{T}}\mathbf{F_y}\mathbf{V}\right\}\mathrm{e}^{+i\mathbf{h}t}\right], \tag{10.14}$$

with

Note that $\mathbf{V}^\mathrm{T}\mathbf{A}\mathbf{V}$ transforms a matrix $\mathbf{A}$ from the $\alpha_\mathrm{I}\alpha_\mathrm{S}$, $\alpha_\mathrm{I}\beta_\mathrm{S}$, ... basis into the eigenbasis of $\mathbf{H}$.

$$\mathbf{V}^\mathrm{T}\mathbf{F_y}\mathbf{V} = \frac{i}{2}\begin{pmatrix} 0 & -(c+s) & -(c-s) & 0 \\ (c+s) & 0 & 0 & -(c+s) \\ (c-s) & 0 & 0 & -(c-s) \\ 0 & (c+s) & (c-s) & 0 \end{pmatrix}. \tag{10.15}$$

Thus,

$$\begin{aligned} \langle \hat{F}_y \rangle = & -\tfrac{1}{2}\cos\left[2\pi\left(\nu - \frac{\varepsilon}{2} + \frac{J}{2}\right)t\right](1+\sin 2\theta) \\ & -\tfrac{1}{2}\cos\left[2\pi\left(\nu - \frac{\varepsilon}{2} - \frac{J}{2}\right)t\right](1-\sin 2\theta) \\ & -\tfrac{1}{2}\cos\left[2\pi\left(\nu + \frac{\varepsilon}{2} + \frac{J}{2}\right)t\right](1-\sin 2\theta) \\ & -\tfrac{1}{2}\cos\left[2\pi\left(\nu + \frac{\varepsilon}{2} - \frac{J}{2}\right)t\right](1+\sin 2\theta). \end{aligned} \tag{10.16}$$

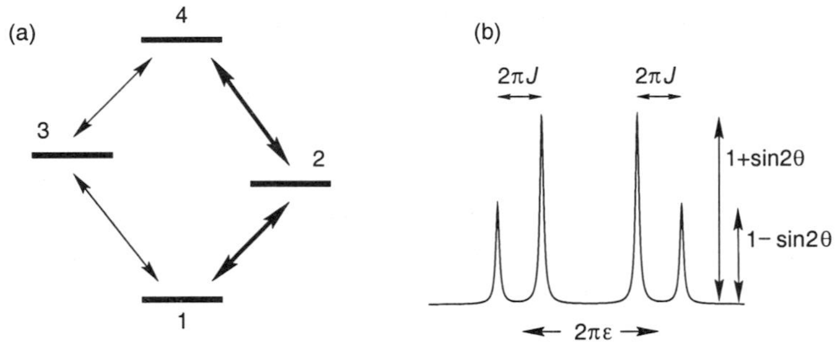

Fig. 10.1 Energy levels (a) and schematic spectrum (b) for a strongly coupled pair of spin-1/2 nuclei. The thick and thin arrows indicate the more allowed and less allowed NMR transitions, respectively.

If J is negative, the pairs of lines swap positions *and* amplitudes so that the spectrum looks exactly the same.

The origin of these strong coupling effects is discussed qualitatively in OCP 32, Sections 3.4 and 3.5.

Equation 10.16 shows that the free induction decay comprises two lines with amplitude proportional to $1 + \sin 2\theta = 1 + J/\varepsilon$, at frequencies $2\pi\nu \pm \pi(\varepsilon - J)$, and two with amplitude proportional to $1 - \sin 2\theta = 1 - J/\varepsilon$, at frequencies $2\pi\nu \pm \pi(\varepsilon + J)$. If J is positive, the former are the stronger, inner lines and the latter the weaker, outer lines of the familiar NMR spectrum of a strongly coupled pair of protons (Fig. 10.1). In the weak coupling limit ($\varepsilon = \delta$, $\sin 2\theta = 0$), the four amplitudes become identical and the frequencies correspond to a pair of doublets, centred at the chemical shift positions, Ω_I and Ω_S, with splitting $2\pi J$ (cf. Eqn 9.23). The physics behind the line shifts and intensity changes predicted for strong coupling lies in the mixing of the $|\alpha_\mathrm{I}\beta_\mathrm{S}\rangle$ and $|\beta_\mathrm{I}\alpha_\mathrm{S}\rangle$ states (Eqn 10.2) and the consequent modification of the NMR transition probabilities. The outer lines become 'less allowed', or 'more forbidden', while the inner lines become 'more allowed'. In the extreme limit of strong coupling, i.e. for two equivalent spins, the outer lines are completely forbidden and have no intensity, and the inner lines coincide (see Section 9.6).

10.3 Spin echoes

We now consider the echo sequence $90°_x - \tau - 180°_x - \tau$ in which the two pulses have the *same* phase. The only difference the phase of the second pulse makes to the outcome of the experiment is that the echo is formed along the $+y$ axis rather than the $-y$ axis, as was found to be the case for $90°_x - \tau - 180°_y - \tau$ (Sections 1.7 and 3.4).

The echo signal is given by the trace of $\mathbf{F_y}$ multiplied by $\rho(2\tau)$, which is obtained by allowing the density operator to evolve sequentially during the four parts of the experiment:

$$\langle \hat{F}_y \rangle = \mathrm{Tr}\Big[\mathbf{F_y} e^{-i\mathbf{H}_{IS}\tau} e^{-i\pi \mathbf{F_x}} e^{-i\mathbf{H}_{IS}\tau} e^{-i\frac{1}{2}\pi \mathbf{F_x}} \times \rho(0) e^{+i\frac{1}{2}\pi \mathbf{F_x}} e^{+i\mathbf{H}_{IS}\tau} e^{+i\pi \mathbf{F_x}} e^{+i\mathbf{H}_{IS}\tau}\Big]. \tag{10.17}$$

Following the same procedure as in the previous section, we obtain

$$\langle \hat{F}_y \rangle = \mathrm{Tr}\Big[\left(\mathbf{V}^{\mathsf{T}}\mathbf{F_y}\mathbf{V}\right) e^{-i\mathbf{h}\tau}\left(\mathbf{V}^{\mathsf{T}} e^{-i\pi\mathbf{F_x}}\mathbf{V}\right) e^{-i\mathbf{h}\tau} \times \mathbf{V}^{\mathsf{T}}\left(-\mathbf{F_y}\right)\mathbf{V}\; e^{+i\mathbf{h}\tau}\left(\mathbf{V}^{\mathsf{T}} e^{+i\pi\mathbf{F_x}}\mathbf{V}\right) e^{+i\mathbf{h}\tau}\Big]. \tag{10.18}$$

Most of the work has already been done; the only part of Eqn 10.18 that has not been determined before is the 180° pulse. The exponential matrix $\exp(\pm i\pi\mathbf{F_x})$ can be obtained in the same way as was $\exp(\pm i\pi\mathbf{F_x}/2)$, or more simply, just by squaring $\exp(\pm i\pi\mathbf{F_x}/2)$:

$$\mathbf{V}^{\mathsf{T}} e^{\pm i\pi\mathbf{F_x}}\mathbf{V} = \mathbf{V}^{\mathsf{T}} e^{\pm i\frac{1}{2}\pi\mathbf{F_x}} e^{\pm i\frac{1}{2}\pi\mathbf{F_x}}\mathbf{V} = \begin{pmatrix} 0 & 0 & 0 & -1 \\ 0 & -2sc & 1-2c^2 & 0 \\ 0 & 1-2c^2 & +2sc & 0 \\ -1 & 0 & 0 & 0 \end{pmatrix}. \tag{10.19}$$

The matrix multiplications here are rather unpleasant if done by hand. Computer programs such as Mathematica and Maple are highly recommended. A good strategy is to proceed in stages: first multiply out the products in parentheses in Eqn 10.18, simplify the resulting matrices, multiply them, simplify again, and then take the trace. There inevitably comes a time, however, when the analytical results of such calculations are too complex to be of much use, or when analytical solutions are simply not possible. This often occurs when the Hamiltonian contains blocks that are larger than 2 × 2, which cannot easily be diagonalized. In such cases, the calculation is better done numerically using methods similar to those outlined above.

After a lot of matrix multiplication, one comes to the result:

$$\langle \hat{F}_y \rangle = 2\cos^2 2\theta \cos 2\pi J\tau - \sin 2\theta(1-\sin 2\theta)\cos(2\pi[J+\varepsilon]\tau) + \sin 2\theta(1+\sin 2\theta)\cos(2\pi[J-\varepsilon]\tau). \tag{10.20}$$

Clearly, the echo modulation for a strongly coupled pair of spins is more complicated than in the weak coupling limit. In place of the $2\cos 2\pi J\tau$ modulation, to which Eqn 10.20 reduces when $\varepsilon \approx \delta >> J$, we have three frequencies with different amplitudes that depend on ratio of J/δ, i.e. the strength of the coupling. Simple behaviour is also regained in the limit that the two spins are equivalent: when $\delta = 0$ and $\varepsilon \approx J$, one finds that $\langle F_y \rangle = 2$, that is, no echo modulation at all, as could have been anticipated from section

9.6. It is also easy to show that there is no echo along the x axis, however weak or strong the coupling.

Bibliography

Part A

Claridge, T. D. W. (1999) *High-Resolution NMR Techniques in Organic Chemistry*. Pergamon Press, Oxford.

Clore, G. M. and Gronenborn, A. M. (1994). Multidimensional heteronuclear nuclear magnetic resonance of proteins. *Methods Enzymol.* **239**, 349–363.

Derome, A. E. (1987). *Modern NMR Techniques for Chemistry Research*. Pergamon Press, Oxford.

Ernst, R. R., Bodenhausen, G. and Wokaun, A. (1987). *Principles of Nuclear Magnetic Resonance in One and Two Dimensions*. Clarendon Press, Oxford.

Freeman, R. (1997). *A Handbook of Nuclear Magnetic Resonance*, 2nd edn. Addison Wesley, Harlow.

Freeman, R. (1997). *Spin Choreography*. Spektrum, Oxford.

Harris, R. K. (1986). *Nuclear Magnetic Resonance Spectroscopy*. Pitman, London.

Hoch, J. C. and Stern, A. S. (1996). *NMR Data Processing*. Wiley-Liss Inc., New York.

Homans, S. W. (1992). *A Dictionary of Concepts in NMR*. Clarendon Press, Oxford.

Hore, P. J. (1995). *Nuclear Magnetic Resonance*. Oxford Chemistry Primers 32. Oxford University Press, Oxford.

Keeler, J., Clowes, R. T., Davis, A. L. and Laue, E. (1994). Pulsed-field gradients: theory and practice. *Methods Enzymol.* **239**, 145–207.

Sanders, J. K. M. and Hunter, B. K. *Modern NMR Spectroscopy*. Oxford University Press, Oxford.

Sørensen, O. W., Eich, G. W., Levitt, M. H., Bodenhausen, G. and Ernst, R. R. (1983). Product operator formalism for the description of NMR pulse experiments. *Progr. NMR Spectrosc.* **16,** 163–192.

Part B

Atkins, P. W. (1983). *Molecular Quantum Mechanics*, 2nd edn. Oxford University Press, Oxford.

Ernst, R. R., Bodenhausen, G. and Wokaun, A. (1987). *Principles of Nuclear Magnetic Resonance in One and Two Dimensions*. Clarendon Press, Oxford.

Goldman, M. (1988). *Quantum Description of High-Resolution NMR in Liquids*. Clarendon Press, Oxford.

Green, N. J. B. (1997). *Quantum Mechanics 1: Foundations*. Oxford Chemistry Primers 48, Oxford University Press, Oxford.

Green, N. J. B. (1998). *Quantum Mechanics 2: The Toolkit*. Oxford Chemistry Primers 65, Oxford University Press, Oxford.

Mathematica: http://www.wri.com/

Maple: http://www.maplesoft.com/

Munowitz, M. (1988). *Coherence and NMR*. John Wiley, Chichester.

Press, W. H., Teukolsky, S. A., Vetterling, W. T. and Flannery, B. P. (1992). *Numerical Recipes in C. The Art of Scientific Computing*, 2nd edn. Cambridge University Press, Cambridge.

Index